Tatyana Purtova

**RFMEMS Periodic Structures: Modelling, Components and Circuits**

Tatyana Purtova

# RFMEMS Periodic Structures: Modelling, Components and Circuits

theory and experiment

Südwestdeutscher Verlag für Hochschulschriften

**Impressum / Imprint**

Bibliografische Information der Deutschen Nationalbibliothek: Die Deutsche Nationalbibliothek verzeichnet diese Publikation in der Deutschen Nationalbibliografie; detaillierte bibliografische Daten sind im Internet über http://dnb.d-nb.de abrufbar.

Alle in diesem Buch genannten Marken und Produktnamen unterliegen warenzeichen-, marken- oder patentrechtlichem Schutz bzw. sind Warenzeichen oder eingetragene Warenzeichen der jeweiligen Inhaber. Die Wiedergabe von Marken, Produktnamen, Gebrauchsnamen, Handelsnamen, Warenbezeichnungen u.s.w. in diesem Werk berechtigt auch ohne besondere Kennzeichnung nicht zu der Annahme, dass solche Namen im Sinne der Warenzeichen- und Markenschutzgesetzgebung als frei zu betrachten wären und daher von jedermann benutzt werden dürften.

Bibliographic information published by the Deutsche Nationalbibliothek: The Deutsche Nationalbibliothek lists this publication in the Deutsche Nationalbibliografie; detailed bibliographic data are available in the Internet at http://dnb.d-nb.de.

Any brand names and product names mentioned in this book are subject to trademark, brand or patent protection and are trademarks or registered trademarks of their respective holders. The use of brand names, product names, common names, trade names, product descriptions etc. even without a particular marking in this works is in no way to be construed to mean that such names may be regarded as unrestricted in respect of trademark and brand protection legislation and could thus be used by anyone.

Coverbild / Cover image: www.ingimage.com

Verlag / Publisher:
Südwestdeutscher Verlag für Hochschulschriften
ist ein Imprint der / is a trademark of
AV Akademikerverlag GmbH & Co. KG
Heinrich-Böcking-Str. 6-8, 66121 Saarbrücken, Deutschland / Germany
Email: info@svh-verlag.de

Herstellung: siehe letzte Seite /
Printed at: see last page
**ISBN: 978-3-8381-3451-2**

Zugl. / Approved by: Ulm, Universität, Diss., 2012

Copyright © 2012 AV Akademikerverlag GmbH & Co. KG
Alle Rechte vorbehalten. / All rights reserved. Saarbrücken 2012

# Acknowledgement

First of all, I would like to thank my supervisor, Professor Dr.-Ing. Hermann Schumacher for his support and motivation throughout my work in his research group. Besides, I would like to thank all cooperation partners in the research projects AMICOM and RFPlatform, in particular Dr. Thomas Lisec, Dr. Christoph Huth and Dr. Joachim Janes from the Fraunhofer Institute of Silicon Technology (ISiT), Dr. Flavio Giacomozzi from the Fondazione Bruno Kessler (FBK), Dr. Tauno Vähä-Heikkilä from the VTT Millilab, Dr. Montserrat Fernandez-Bolanos Badia from EPFL, Dr. Paola Farinelli, Dr. Luca Marcaccioli and Dr. Benedetta Mencagli from the University of Perugia and Dr. Vitaly Leus from Technion. I would like to thank Professor Dr.-Ing. Ingmar Kallfass from the Karlsruhe Institute of Technology and Fraunhofer Institute for Applied Physics (IAF) for the cooperation on the RFMEMS-mHEMT integration. During my work at the Ulm University I was lucky to supervise several great students, who I thank for their questions and ideas, which often helped me to look at things from new perspectives: M. Zakir Alam, Naser Pour Aryan, Roberto Astete Mora, Debalina Chaterjee, Chao Chu, Wenfen Liu, M. Ayaaz Qureshi and Cheng Yuan (in alphabetic order). Special thanks go to my current and former university colleagues for inspiring discussions and to some of them also for the proof-reading of this thesis: Nabil Alomari, Dr. Christoph Bromberger, Dr. Sebastien Chartier, Dr. Jochen Dederer, Till Feger, Xiaolei Gai, Dr. Michael Hosch, Dayang Lin, Gang Liu, Peter Lohmiller, Yakiv Men, Giuseppe Oliveri, Guangwen Qu, Dr. Bernd Schleicher, Ferdinand Sigloch, Ursula Spitzberg, Yinmei Su, Dr. Andreas Trasser, Dr. Vaclav Valenta and Ahmet Cagri Ulusoy (in alphabetic order). I would like to appreciate that I had a chance to work and be friends with Dr. Thomas Haag and Miroslav Masar, who unexpectedly passed away in April 2011 and July 2011, they will always be missed. Most of all, I would like to thank my friends, my parents and my husband Sergey for their unconditional support and encouragement.

# Contents

**Abstract**     13

**List of Abbreviations and Symbols**     15

**1 Introduction**     19
     1.1 Motivation . . . . . . . . . . . . . . . . . . . . . . . . . . . . . . . . 19
     1.2 Objectives of the Dissertation . . . . . . . . . . . . . . . . . . . . . 21
     1.3 Dissertation Outline . . . . . . . . . . . . . . . . . . . . . . . . . . 22

# I Developed RFMEMS Components     23

**2 Introduction**     25

**3 Theoretical Background of RFMEMS**     27
     3.1 RFMEMS Actuation Mechanisms . . . . . . . . . . . . . . . . . . . 27
     3.2 Electromechanical Analysis of Electrostatic MEMS . . . . . . . . . 30
         3.2.1 Static Pull-in . . . . . . . . . . . . . . . . . . . . . . . . . . 30
         3.2.2 Dynamic Pull-in . . . . . . . . . . . . . . . . . . . . . . . . 32
         3.2.3 Switching Time . . . . . . . . . . . . . . . . . . . . . . . . 35
         3.2.4 Release Time . . . . . . . . . . . . . . . . . . . . . . . . . 36
     3.3 Electromagnetic Analysis of Electrostatic MEMS . . . . . . . . . . 37
         3.3.1 Shunt Capacitive Switches and Switchable Capacitors . . . . . . . . . 37

|  |  |  |  |
|---|---|---|---|
|  | 3.3.2 | Series Switches with Metal-to-Metal Contact . . . . . . . . . . . . . | 40 |

## 4 Components in ISiT RFMEMS Technology — 43

- 4.1 ISiT RFMEMS Fabrication Process . . . . . . . . . . . . . . . . . . . . . . . 43
  - 4.1.1 ISiT RFMEMS Fabrication Steps . . . . . . . . . . . . . . . . . . 43
  - 4.1.2 ISiT Standard Switch . . . . . . . . . . . . . . . . . . . . . . . . . 45
  - 4.1.3 Microwave Losses of Oxidized High-resistivity Silicon . . . . . . . . 48
- 4.2 Shunt Capacitive SPST Switch . . . . . . . . . . . . . . . . . . . . . . . . . 48
  - 4.2.1 Design . . . . . . . . . . . . . . . . . . . . . . . . . . . . . . . . . 48
  - 4.2.2 Electromagnetic Characterisation . . . . . . . . . . . . . . . . . . . 52
  - 4.2.3 Electromechanical Characterisation . . . . . . . . . . . . . . . . . . 53
- 4.3 Shunt Capacitive SPST Switch for a 26 GHz Phase Shifter . . . . . . . . . . 55
  - 4.3.1 Design . . . . . . . . . . . . . . . . . . . . . . . . . . . . . . . . . 55
  - 4.3.2 Electromagnetic Characterisation . . . . . . . . . . . . . . . . . . . 58
  - 4.3.3 Electromechanical Characterisation . . . . . . . . . . . . . . . . . . 58
- 4.4 Tri-State Shunt Switchable Capacitor . . . . . . . . . . . . . . . . . . . . . . 59
  - 4.4.1 Design . . . . . . . . . . . . . . . . . . . . . . . . . . . . . . . . . 59
  - 4.4.2 Electromagnetic Characterisation . . . . . . . . . . . . . . . . . . . 61
  - 4.4.3 Electromechanical Characterisation . . . . . . . . . . . . . . . . . . 61
- 4.5 Two-state Shunt Switchable Capacitor . . . . . . . . . . . . . . . . . . . . . 63
  - 4.5.1 Design . . . . . . . . . . . . . . . . . . . . . . . . . . . . . . . . . 63
  - 4.5.2 Electromagnetic Characterisation . . . . . . . . . . . . . . . . . . . 64
  - 4.5.3 Electromechanical Characterisation . . . . . . . . . . . . . . . . . . 64
- 4.6 Switchable Series Capacitor . . . . . . . . . . . . . . . . . . . . . . . . . . . 65
  - 4.6.1 Design . . . . . . . . . . . . . . . . . . . . . . . . . . . . . . . . . 65
  - 4.6.2 Electromagnetic Characterisation . . . . . . . . . . . . . . . . . . . 66
  - 4.6.3 Electromechanical Characterisation . . . . . . . . . . . . . . . . . . 67

| | | |
|---|---|---|
| 4.7 | Cantilever-based Shunt Switchable Capacitor | 69 |
| | 4.7.1 Design | 69 |
| | 4.7.2 Electromagnetic Characterisation | 70 |
| | 4.7.3 Electromechanical Characterisation | 70 |

## 5 Components in FBK RF-MEMS Technology — 73

| | | |
|---|---|---|
| 5.1 | FBK RFMEMS Fabrication Process | 73 |
| 5.2 | Series Metal-to-Metal Contact Switch | 75 |
| | 5.2.1 Design | 75 |
| | 5.2.2 Electromagnetic Characterisation | 75 |
| | 5.2.3 Electromechanical Characterisation | 76 |
| 5.3 | RFMEMS-Switchable Series Inductance using Defected Ground Structures | 78 |
| | 5.3.1 Design | 78 |
| | 5.3.2 Electromagnetic Characterisation | 79 |
| | 5.3.3 Electromechanical Characterisation | 80 |
| 5.4 | Shunt Switchable Capacitor | 82 |
| | 5.4.1 Design | 82 |
| | 5.4.2 Electromagnetic Characterisation | 82 |
| | 5.4.3 Electromechanical Characterisation | 83 |

# II Modelling of RFMEMS Periodic Structures — 87

## 6 Introduction — 89

## 7 Theoretical Background — 93

| | | |
|---|---|---|
| 7.1 | Space Harmonics Analysis of General 1-D Periodic Structures | 93 |
| 7.2 | ABCD-Analysis of General 1-D Periodic Structures | 96 |
| | 7.2.1 Periodic Structures with Finite Number of Sections | 96 |

|       |       |                                                                      |     |
|-------|-------|----------------------------------------------------------------------|-----|
|       | 7.2.2 | Effective Propagation Constant                                       | 98  |
|       | 7.2.3 | Pass- and Stop-bands                                                 | 99  |
|       | 7.2.4 | Bloch Impedance                                                      | 100 |
|       | 7.2.5 | Special Case of Lossless and Symmetrical Unit Cell                   | 100 |

# 8  Advanced Modelling of RFMEMS Periodic Structures   103

| | | | |
|---|---|---|---|
| 8.1 | Lossless Periodic Structure with Shunt Capacitive Loading | | 103 |
| | 8.1.1 | ABCD-Parameters of a Unit Cell | 103 |
| | 8.1.2 | Effective Propagation Constant | 104 |
| | 8.1.3 | Bloch Impedance | 107 |
| | 8.1.4 | Cut-off Frequencies of Pass- and Stopbands | 107 |
| | 8.1.5 | Experimental Validation of the Model | 110 |
| 8.2 | Lossless Periodic Structure with Series Inductive Loading | | 117 |
| | 8.2.1 | ABCD-Parameters of a Unit Cell | 117 |
| | 8.2.2 | Effective Propagation Constant | 118 |
| | 8.2.3 | Bloch Impedance | 118 |
| | 8.2.4 | Cut-off Frequencies of Pass- and Stopbands | 120 |
| 8.3 | Lossless Periodic Structure with Series Inductive and Shunt Capacitive Loading | | 120 |
| | 8.3.1 | ABCD-Parameters of a Unit Cell | 120 |
| | 8.3.2 | Bloch Impedance | 122 |
| | 8.3.3 | Effective Propagation Constant | 123 |
| | 8.3.4 | Cut-off Frequencies of Pass- and Stopbands | 125 |
| | 8.3.5 | Experimental Validation of the Model | 125 |

# III  Circuits Utilizing RFMEMS Periodic Structures   129

# 9  Distributed RFMEMS Phase Shifters   131

    9.1  22 GHz DMTL Phase Shifter with Capacitive and Inductive Tuning       132

    9.2   26 GHz One-port Reflection-Type Phase Shifter with Shunt Capacitive Switches 134

          9.2.1   26 GHz Beam-steerable Reflectarray . . . . . . . . . . . . . . . . . 135

          9.2.2   Reflection-Type RFMEMS Phase Shifter . . . . . . . . . . . . . . . 136

    9.3   64 GHz DMTL Phase Shifter with Capacitive Tuning . . . . . . . . . . . . . 139

    9.4   82 GHz DMTL Phase Shifter with Capacitive Tuning . . . . . . . . . . . . . 142

## 10  Distributed RFMEMS Impedance Tuners                        145

    10.1  Ka-Band Impedance Tuner with Capacitive and Inductive Tuning . . . . . . . 145

    10.2  W-band Impedance Tuner with Capacitive Tuning: Design 1 . . . . . . . . . 149

    10.3  W-band Impedance Tuner with Capacitive Tuning: Design 2 . . . . . . . . . 151

# Conclusions and Outlook                                                155

## 11  Conclusions                                                                        155

## 12  Outlook                                                                                    157

    12.1  Millimeter-wave RFMEMS and Their Monolithic MMIC Integration . . . . . . . 157

    12.2  Monolithic Integration of RFMEMS into the mHEMT Process of Fraunhofer IAF 158

# Appendix                                                                                            165

**A**  **Approximate Solutions of Equation** $\tan(x) = -\xi x$                 **165**

**B**  **Approximate Solution of Equation** $\cot(x) = +\xi x$                   **169**

**Bibliography**                                                                                         **182**

**List of Publications**                                                               **182**

# Abstract

This thesis deals with RFMEMS-reconfigurable periodic structures. In general, periodic structures are obtained by cascading a non-uniform unit cell a large number of times and are utilized in many applications. Reconfigurable periodic structures are widely used as phase shifters, tunable filters, tunable impedance tuners. A periodic structure can be made reconfigurable by incorporating some adjustable elements into its unit cell. Due to the advancement of micromachining technologies, RFMEMS tunable or switchable capacitors became an attractive way to achieve reconfigurability. As compared to their semiconductor and ferroelectric counterparts, RFMEMS components offer lower loss and higher linearity even at elevated frequencies. These properties are especially advantageous to reduce losses of phase shifters, improve power-handling capability of load-pull impedance tuners and extend impedance coverage of tuners used in noise parameter measurements, to name but a few examples. However, most of RFMEMS technologies are not yet mature and issues like reliability, packaging and yield are yet to be conclusively solved at the technology side. This work is focused mainly on the modelling and applications and less on the technology development.

The first part of the thesis covers the design of diverse RFMEMS components to be used in unit cells of periodic structures. The developed components include two- and three-state shunt and series switchable capacitors as well as an RFMEMS-switchable inductor. The components were designed using the RFMEMS technology at Fraunhofer Institute for Silicon Technology (ISiT) in Germany and at Fondazione Bruno Kessler (FBK) in Italy.

By now, a lot of work has been conducted to develop distributed MEMS transmission line (DMTL) phase shifters (e.g. [1, 2]) and impedance tuners (e.g. [3]), and advantages of RFMEMS periodic structures have been clearly demonstrated. The circuit design procedure often follows the one proposed by S. Barker in [1], which is based on lumped approximation and works well up to about one-third of the cut-off frequency of the first passband of the periodic structure. Although this approach is relatively straight-forward and simple, its accuracy strongly degrades for higher frequencies, exceeding one-third of the cut-off frequency. On the other hand, phase and impedance variations increase as the frequency gets closer to the passband-stopband transition, which is advantageous for phase shifters and impedance tuners respectively. In other words, circuits operating close to the cut-off frequency can potentially result in larger phase shift or impedance variation for the same loss and area.

Thus, the second part of the thesis is devoted to the development of a broadband dispersion model of distributed MEMS transmission lines. The model extends the results derived in [4]. The emphasis is on right-handed structures including the commonly-used shunt capacitive loading as well as series inductive loading and combination of both shunt capacitive and series inductive loading. The latter can potentially overcome the inherent trade-off between impedance mismatch and phase shift. Simple, accurate and frequency-dependent equations for the propagation constant and equivalent characteristic impedance are obtained by adopting the "loading factor" concept used in [5]. Besides, accurate equations for cut-off frequencies of all pass- and stopbands are derived for the cases of shunt capacitive or series inductive loading.

The third part of the thesis presents RF-MEMS phase shifters and impedance tuners based on the periodic structures utilizing the components developed in the first part. The part covers Ka-, V- and W-band phase shifters with capacitive and capacitive-inductive tuning. Operation frequency of phase shifters was extended from the recommended 30 % [6] to 70 % of the cut-off of the first passband. Besides, Ka- and W-band impedance tuners also with capacitive as well as capacitive-inductive tuning were developed. Good impedance coverage at V- and W-bands was obtained, making the circuits suitable for load-pull and noise parameter measurement systems.

# List of Abbreviations and Symbols

## List of Abbreviations

| BiCMOS | Bipolar Complementary Metal Oxide Semiconductor |
|---|---|
| CMOS | Complementary Metal Oxide Semiconductor |
| CMP | Chemical Mechanical Polishing |
| CPW | Coplanar Waveguide |
| CRLH | Composite Right/Left Handed |
| DC | Direct Current |
| DGS | Defected Ground Structure |
| DMTL | Distributed MEMS Transmission Line |
| EM | Electromagnetic |
| FET | Field Effect Transistor |
| FBK | Fondazione Bruno Kessler |
| GSG | Ground-Signal-Ground |
| HEMT | High Electron Mobility Transistor |
| IAF | Fraunhofer Institute for Applied Solid-State Physics |
| ISiT | Fraunhofer Institute for Silicon Technology |
| KIT | Karlsruhe Institute of Technology |
| LDV | Laser Doppler Vibrometer |
| LH | Left-Handed (Transmission) |
| MEMS | Micro Electro-Mechanical System |
| mHEMT | metamorphic High Electron Mobility Transistor |
| MIM | Metal-Insulator-Metal |

| | |
|---|---|
| MMIC | Micro-/Millimeter-wave Monolithic Integrated Circuit |
| MOS | Metal Oxide Semiconductor |
| MPW | Multi Project Wafer(-Run) |
| PECVD | Plasma-Enhanced Chemical Vapour Deposition |
| PZT | Lead Zirconate Titanate |
| RFCMOS | Radio Frequency Complementary Metal Oxide Semiconductor |
| RFMEMS | Radio Frequency Micro Electro-Mechanical System |
| RH | Right-Handed (Transmission) |
| SEM | Scanning Electron Microscope |
| SOLT | Short-Open-Load-Thru |
| SPDT | Single Pole Double Throw |
| SPST | Single Pole Single Throw |
| SRR | Split Ring Resonator |
| T/R | Transmit/Receive |

# List of Symbols

| | |
|---|---|
| $\alpha$ | attenuation constant of a loaded transmission line |
| $\beta$ | phase constant of a loaded transmission line |
| $\beta_0$ | phase constant of an unloaded transmission line |
| $\gamma$ | propagation constant of a loaded transmission line |
| $\delta_{1,2}$ | coefficients to express $V$ as a function of $V_{dyn.pull-in}$, see Eq. 3.14 |
| $\epsilon_0$ | vacuum permittivity |
| $\epsilon_r$ | relative permittivity |
| $\lambda$ | wavelength in medium |
| $\lambda_0$ | wavelength in free-space |
| $\phi$ | phase delay |
| $\eta$ | damping coefficient |
| $\omega$ | angular frequency |
| $\omega_0$ | mechanical resonance frequency without damping |

| | | |
|---|---|---|
| $\omega_d$ | mechanical resonance frequency with damping | |
| $a$ | actuation area | |
| $A$ | A-parameter of ABCD parameters | |
| $B$ | B-parameter of ABCD parameters | |
| $c$ | constant | |
| $C$ | MEMS capacitance and C-parameter of ABCD parameters | |
| $C'$ | capacitance per unit length | |
| $C_{up}$ | MEMS capacitance in the up-state | |
| $C_{down}$ | MEMS capacitance in the down state | |
| $d$ | unit cell (transmission line) length | |
| $D$ | D-parameter of ABCD parameters | |
| $E$ | electric field | |
| $f$ | frequency | |
| $f_{res}$ | resonance frequency | |
| $F_{el}$ | electrostatic force | |
| $F_{spring}$ | spring force | |
| $g_0$ | zero-voltage air-gap | |
| $I$ | current | |
| $j$ | imaginary unit | |
| $k$ | spring constant | |
| $L$ | MEMS inductance | |
| $L'$ | inductance per unit length | |
| $LF_C$ | capacitive loading factor | |
| $LF_L$ | inductive loading factor | |
| $m$ | spring mass | |
| $N$ | number of unit cells in a periodic structure | |
| $R$ | MEMS resistance | |
| $S$ | S-parameter | |
| $t_d$ | dielectric thickness | |
| $t_s$ | MEMS switching time | |

| $V$ | voltage |
|---|---|
| $v$ | velocity |
| $x$ | deflection |
| $x_{eq}$ | deflection at the equilibrium position |
| $x_{pull-in}$ | deflection at pull-in |
| $x_s$ | stagnation deflection |
| $Z_0$ | characteristic impedance of transmission line |
| $Z_b$ | Bloch impedance |

# Chapter 1

# Introduction

## 1.1 Motivation

This thesis deals with RFMEMS-tunable periodic structures. In general, periodic structures are obtained by cascading some identical non-uniform elements (typically sections of transmission lines with lumped reactances) a large number of times. The periodicity can be present in one-, two- or three dimensions. Unlike uniform materials, whose electrical properties typically do not change much with frequency, properties of periodic structures are highly frequency-dependent. Most importantly, all periodic structures possess [7]:

- frequency passbands, where waves propagate freely through the structure
- frequency stopbands, where wave propagation is not possible

This remarkable behaviour has been utilized in numerous applications dealing both with radiated and guided waves. A large group of periodic structures operate in the frequency passbands, while frequency stopbands are undesirable. The most important examples of such circuits are right-/left-handed slow-wave structures and phased- or reflect-array antennas. Besides, some applications rely on the existence of frequency stopbands, e.g. photonic crystals and electromagnetic band-gap structures.

The focus of this work is on tunable periodic structures. Tunability can be obtained by incorporating some elements, whose electrical properties can be changed according to a control signal. The most important applications of such structures are switchable multi-band circuits and inherently "to-be-adjusted" circuits like phase shifters and impedance tuners. Also, tunable metamaterials including composite-right-left-handed (CRLH) transmission lines received a lot of attention recently.

Often, periodic structure are made tunable by means of semiconductor varactors. A large number of varactor-tunable circuits based on CRLH was reported. In [8], CRLH transmission lines with series and shunt varactor diodes were implemented to realize a tunable leaky-wave antenna. In [9], tunable 0°-phase shift transmission line based on CRLH lines was designed. CRLH-based phase shifters were presented in [10–12] and an impedance tuner in [13]. Finally, [14, 15] described tunable delay lines , while a tunable directional coupler was reported in [16]. Properties of periodic structures can also be adjusted with ferroelectric varactors. In [17, 18], phase shifters based on CRLH transmission lines with shunt and series ferroelectric varactors were designed. In [19], superconducting tunable CRLH transmission lines were reported.

Another approach to obtain tunable microwave and millimeter-wave periodic structures is by incorporating RFMEMS into them. RFMEMS stands for radio frequency microelectromechanical system, yet this term has become a general expression for microelectromechanical devices for microwave- and millimeter-wave applications. The most common RFMEMS devices are switches and varactors. As compared to their semiconductor and ferroelectric counterparts, RFMEMS components offer lower loss and higher linearity even at elevated frequencies. From the periodic structure point of view, these properties are especially advantageous to reduce losses of phase shifters, improve power-handling capability of load-pull impedance tuners and extend impedance coverage of impedance tuners used in noise parameter measurements, to name but a few examples.

By now, a lot of work has been conducted to develop distributed MEMS transmission line (DMTL) phase shifters (e.g. [1, 2]) and impedance tuners (e.g. [3]), and advantages of RFMEMS periodic structures have been clearly demonstrated. The circuit design procedure often follows the one proposed by S. Barker in [1], which is based on lumped approximation and works well up to about one-third of the cut-off frequency of the first passband of the periodic structure. Although this approach is relatively straight-forward and simple, its accuracy strongly degrades for higher frequencies, exceeding one-third of the cut-off frequency.

On the other hand, phase and impedance variations increase as the frequency gets closer to passband-stopband transition, which is advantageous for phase shifters and impedance tuners respectively. In other words, circuits operating close to the cut-off frequency can potentially result in larger phase shift or impedance variation for the same loss and area. Thus, a broadband model valid up to (and potentially beyond) the cut-off frequency of the periodic structure is needed. Besides, operation of circuits close to the cut-off frequency requires a new design methodology, which takes into account the periodic nature of the circuit and the properties arising due to pass- and stopband of periodic structures.

Most of the reported circuits utilize periodic structures containing switchable or tunable RFMEMS capacitors. However, purely-capacitive periodic structures suffer from the inherent trade-off between impedance mismatch and phase shift, which can strongly deteriorate the performance of DMTL phase shifters. Although some solutions to reduce the impedance mismatch have been proposed ( [4, 20, 21]), they do not eliminate the inherent mismatch trade-off and some approaches require complex control networks.

To overcome this problem, Chapter 8.3 of this thesis suggests that the phase shift-mismatch trade-off can be eliminated completely by using both switchable shunt capacitors and series inductors, thus the phase shifter losses can be significantly reduced. This also is proven analytically. However, RFMEMS-switchable or tunable inductors are not readily available and very few results on have been reported so far [22, 23]. Thus, there is a strong demand for low-loss tunable or switchable inductors for microwave and millimeter-wave frequencies.

## 1.2 Objectives of the Dissertation

This work is devoted to tunable one-dimensional periodic structures for guided waves with the emphasis on their applications as distributed RFMEMS phase shifters and distributed RFMEMS impedance tuners. Objectives of the thesis can be summarized as follows:

- develop RF-MEMS components suitable for application in periodic structures

- derive closed-form equations for diverse RFMEMS periodic structures, which can be used for practical designs of periodic structures

- develop practical circuits based on periodic structures, including new topologies with inductive tuning

## 1.3 Dissertation Outline

This thesis is divided into three parts. Each part starts with a concise review of the related theoretical background, followed by design methodology and experimental results.

**Part 1** describes developed RFMEMS components in ISiT and FBK RFMEMS technologies. Diverse RFMEMS switches, switchable capacitors and a switchable inductor are presented. The focus is on the component development from the RF-designer point of view, rather than from the technological one. Both electromagnetic and electromechanical analysis are presented.

**Part 2** deals with advanced modelling of RFMEMS periodic structures. The goal is to obtain simple but accurate closed-form equations to be used by RF-designers developing periodic-structure-based circuits. The developed closed-form equations can be used also for general periodic structures without RFMEMS, provided that the structures are low-loss as is the case for RFMEMS. The part starts from general theoretical description of periodic structures. Then, several RFMEMS-switchable periodic structures are studied. In particular, the case of both shunt capacitive and series inductive loading is considered in detail. The derived equations are confirmed experimentally up to 110 GHz.

**Part 3** discusses the realization of phase shifters and impedance tuners based on the RFMEMS periodic structures. The circuits are developed using the RFMEMS components presented in Part 1.

# Part I

# Developed RFMEMS Components

# Chapter 2

# Introduction

Switches and variable capacitors are important building blocks of integrated circuits. Due to the rapid advancement of RFCMOS technologies, MOS switches are gaining in importance, but still are used typically below 6 GHz, and rarely up to 20 GHz (see e. g. [24–26]). For millimetre-wave applications III-V FET or pin-diode switches can be used, but they suffer from increasing losses at high frequencies and limited linearity. Similar arguments apply to semiconductor varactors.

On the contrary, RFMEMS can provide inherently high linearity and low loss even at elevated frequencies and are substrate-independent. A lot of work has been conducted worldwide to develop reliable and reproducible MEMS components. To mention but a few are the series ohmic switches of Radant [27] and Omron [28] as well as cantilever-type capacitive switches of EADS Innovation Works [29–31]. To summarize, the main motivation behind development of RFMEMS components is their potentially superior electromagnetic performance, particularly important at millimeter-wave frequencies, where the performance of their semiconductor counterparts can considerably degrade.

Despite all potential benefits of RFMEMS components, one should not overlook inherent limitations arising due to the mechanical nature. First of all, the switching speed is determined by the mechanical movement of the suspended plate and is typically a few tens of microseconds, which may or may not be a limitation, depending on the intended application. Besides, RFMEMS can have relatively large footprints (few hundreds of micrometers in length and width), mainly due to the large area of actuation electrodes.

Another important issue is reliability of RFMEMS. The main reliability problems is sticking due to dielectric charging in case of electrostatic actuation, moisture or contact welding under hot switching (the latter applies to ohmic contact components). Solutions to these problems are the use of special dielectric layers less prone to charging (e.g. aluminium nitride) or dielectric-less electrodes with mechanical stoppers on the membrane preventing a DC short, hermetic packaging, special contact materials or cold switching. Besides, variation of temperature may lead to deformation of the thin moveable membranes, resulting in a shift in pull-in voltage or different insertion loss or isolation. This can be addressed by appropriate mechanical design of the suspended membrane.

Obtaining high RFMEMS yield is a large technological challenge, since it requires well-controlled and uniform stress in the switch membranes over the whole wafer combined with excellent process stability and reproducibility. As a result, most RFMEMS processes are not yet commercially available and are undergoing a constant development and improvement. Packaging of RFMEMS is another important issue. Reliability, yield and packaging are mainly technological issues and should be solved on the foundry side by choice of suitable materials and adjustments of fabrication parameters. Thus, these issues are touched upon, but their solutions are not primary goals of the work.

This part reports on RFMEMS components designed in Fraunhofer ISiT (Itzehoe, Germany) and FBK (Trento, Italy) technologies. The purpose is to develop switchable capacitors and inductors required for RFMEMS periodic structure applications presented in Part III. Besides, detailed analysis and modelling of switching dynamics rarely reported in the literature is performed. Before description of the designed components, the theoretical background of RFMEMS design is reviewed.

# Chapter 3

# Theoretical Background of RFMEMS

## 3.1 RFMEMS Actuation Mechanisms

Radio frequency microelectromechanical systems (RFMEMS) can be realized in very diverse ways, but all of them can be generally described by an equivalent mass-spring model [32]. Fig. 3.1 shows such a model for the electrostatic case, but other actuation mechanisms can be described in a similar way. The model consists of a pair of plates: (i) one suspended in the air by a spring and (ii) one fixed to a surface. The suspended plate can be moved towards the fixed plate by means of either electrostatic, thermal, electromagnetic or piezoelectric forces. This movement can be utilized for the design of a microelectromechanical switch with either capacitive or ohmic (metal-to-metal) contact. Besides, tunable or switchable capacitors and to a lesser extent switchable inductors can be developed.

Presence of a moveable part is the main distinguishing feature of microelectromechanical systems. It clearly differentiates them from just "micromachined", or membrane-supported, components, where the goal is to either improve quality factors of passives (e.g. spiral inductors [23] or transmission lines [33]) or to increase antenna bandwidth [34]. However, micromachining on its own does not result in adjustable electrical properties of components. In contrast, microelectromechanical systems must contain a moveable part to change their electrical behavior.

Main pros and cons of different RFMEMS actuation mechanisms are summarized in the following. A **thermal** actuator is typically made of a stack of two materials with different temperature expansion coefficients. When heated by passing current through the stack, each layer undergoes a different thermal expansion. This results in a bending moment, which can be utilized for MEMS actuation. Unfortunately, MEMS components with thermal actuation suffer from relatively large response times, must be biased with low-resistance lines which can interfere with RF signals and have large DC power consumption. The latter problem can be mitigated with bistable MEMS designs [35], which consume DC power only during switching and release.

**Electromagnetic** actuation employs moveable ferromagnetic membranes (often made of NiFe) and electromagnets with low-resistance coils to reduce DC power dissipation. When a current is passed through the coil, the generated non-uniform magnetic field can induce large displacements of the membrane with large contact force. However, fabrication of compact, low-loss coils with a large number of windings may be difficult. Besides, these components must be biased with low resistance lines, which may electromagnetically couple with RF signal lines. In addition, required magnetic materials are not part of standard microelectronics fabrication. In addition, this actuation mechanism leads to large DC power consumption.

As the name suggests, **piezoelectric** actuators are covered with piezoelectric materials (often PZT), producing a mechanical force in response to an applied voltage. Since no current is involved, no DC power is consumed. Besides, large deflection and large contact forces are feasible. However, piezoelectric materials and related fabrication techniques are often not available in microelectronics fabrication.

**Electrostatic** actuation relies on an attractive force between two metal plates, when a potential difference is applied between them. The electrostatic actuation is by far the most commonly used due to its negligible DC power consumption, fast switching as compared to other actuation mechanisms (e.g. thermal), possibility of biasing with high-resistivity lines and fabrication processes compatible with standard microelectronics techniques. Electrostatic actuation was used in this work.

Despite all potential benefits of the electrostatic actuation, it typically requires a DC voltage in the order of few tens of volts to create the sufficient driving forces. This leads to the following problems:

- if the required high voltage is not available on-chip, a charge-pump should be used. This requires, however, high breakdown voltage devices, at least as an option, as well as sufficiently robust on-chip capacitors. Additionally, the charge pumps will consume power and chip area. Fortunately, high voltage is required only for a short time during switching. After that the moveable membrane can be held in the down-state with a much smaller voltage, since the metal plates are now very close to each other. Thus, with some control circuitry it is possible to turn off the charge pump immediately after switching, significantly reducing the overall power consumption.

- high electric field strength may cause dielectric "charging", i.e. injection and trapping of charges in the dielectric layer leading to a residual electric field after the actuation voltage is removed. Charging can result in either sticking or repulsion of the MEMS membrane. It is possible to cope with dielectric charging either by special designs avoiding dielectrics in high electric field regions (except air), or technologically by selecting a dielectric material less prone to charging. In this work the latter solution is employed, since the used RFMEMS technologies provide suitable dielectric materials with low tendency to charging (AlN in ISiT technology and thermally-grown $SiO_2$ in FBK technology).

The problem of high actuation voltage could in principle be mitigated by low spring constant designs. However, this is not used in practice and is also avoided in this work for the following two reasons. Firstly, a low spring constant leads to a low restoring force, making the moveable membrane much more prone to sticking, significantly degrading its reliability. Besides, a low spring constant may greatly increase the switching time of MEMS devices (see Eq. 3.14). Thus, to achieve a compromise between high spring constant and a reasonable actuation voltage, it is often required to increase MEMS actuation electrodes, leading to relatively large footprints of few hundreds of micrometers in length and width.

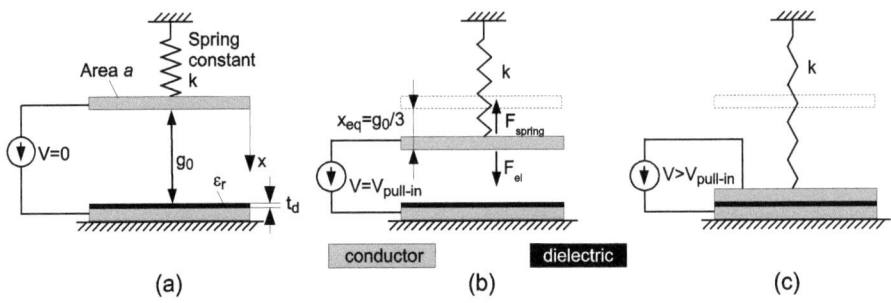

Figure 3.1: Parallel-plate model of an electrostatically-actuated RFMEMS: (a) initial position, (b) at static pull-in, (c) after static pull-in.

## 3.2 Electromechanical Analysis of Electrostatic MEMS

### 3.2.1 Static Pull-in

Static analysis assumes that the actuation voltage is increasing slowly enough, so that at any moment of time the system is in a static equilibrium, i. e. the total force is zero and the suspended plate is not moving. This somewhat artificial approach allowing us to understand important physical processes using a very simple model considered below.

As mentioned above, Fig. 3.1 illustrates a general spring-mass model of an electrostatic RFMEMS. Due to an applied potential difference $V$, charges of different polarity arise on the two plates, generating an attractive electrostatic force, which is utilized for MEMS actuation. The model has one degree-of-freedom, allowing movement only along the x-axis. For brevity, but without loss of physical insight, the following analysis neglects fringing electric fields and bending of the moveable plate, i.e. it assumes an ideal parallel-plate movement [32].

Initially, the two plates are at the same DC-potential (Fig. 3.1a, $V = 0$) and the spring is at equilibrium. As a small potential difference $V$ is applied (Fig. 3.1b), an attractive electrostatic force $F_{el}$ is created between the plates and the suspended plate moves towards the fixed plate. Simultaneously, the spring force $F_{spring}$ increases to oppose this movement.

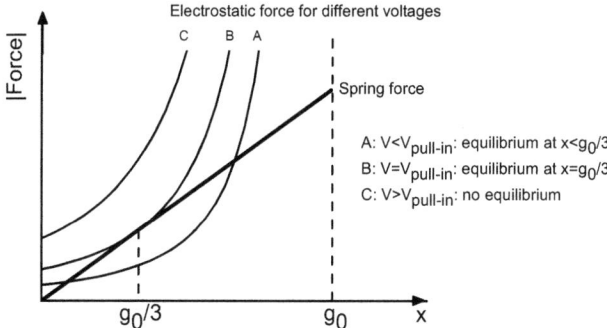

Figure 3.2: Magnitudes of electrostatic and spring forces acting on a suspended plate shown in Fig. 3.1 versus plate displacement $x$.

For most RFMEMS the displacement is small, so that the spring remains linear and the spring force $F_{spring}$ is defined by Hooke's law (Eq. 3.1)[1]. The conductive plates in Fig. 3.1 have an area $a$, are separated from each other by a distance $g_0$, the dielectric thickness is $t_d$ and the suspending spring has a spring constant $k$. So, the electrostatic force $F_{el}$ is given by Eq. 3.2.

$$F_{spring} = -kx \qquad (3.1)$$

$$F_{el} = \frac{1}{2}\frac{\epsilon_0 a V^2}{(g_0 + \frac{t_d}{\epsilon_r} - x)^2} \qquad (3.2)$$

For small $V$ both forces balance each other at an equilibrium position $g_0 - x_{eq}$, obtained by equating their magnitudes: $\mid F_{el} \mid = \mid F_{spring} \mid$. However, above a certain voltage called the "static pull-in voltage", the electrostatic force exceeds the spring force for all displacements. Thus, no equilibrium position exists and the suspended plate moves all the way down onto the fixed plate, as illustrated in Fig. 3.1c. The application of this so-called "pull-in" effect in microelectronics circuits was first reported by Nathanson et al. more than forty years ago in their research on resonant gate transistors [36].

The origin of the static pull-in is explained in Fig. 3.2, where magnitudes of electrostatic and spring forces are qualitatively plotted versus the displacement $x$. Curve $A$ corresponds

---
[1]The spring displacement is much smaller than the not-to-scale Fig. 3.1 shows.

to a small voltage, at which an equilibrium position does exist. Curve $C$ corresponds to a high voltage, when $|F_{el}|$ is always larger than $|F_{spring}|$ and there is no stable point for the moving plate. Curve $B$ illustrates the limiting case, corresponding to the "pull-in" voltage.

The pull-in voltage $V_{pull-in}$ and the corresponding equilibrium position $x_{pull-in}$ are derived from the cross-sectional point of the curve $B$ and the spring force in Fig. 3.2, where both magnitudes and their partial derivatives are equal:

$$\begin{cases} |F_{el}| = |F_{spring}| \\ \dfrac{\partial |F_{el}|}{\partial x} = \dfrac{\partial |F_{spring}|}{\partial x} \end{cases}$$

Then the solutions for $x_{pull-in}$ and $V_{pull-in}$ are:

$$\begin{cases} x_{stat.pull-in} = \dfrac{(g_0 + \frac{t_d}{\epsilon_r})}{3} \\ V_{stat.pull-in} = \sqrt{\dfrac{8k(g_0 + \frac{t_d}{\epsilon_r})^3}{27\epsilon_0 a}} \end{cases} \quad (3.3)$$

Thus, the static pull-in happens, when the equilibrium position $x_{eq}$ equals exactly one-third of the zero-voltage plates separation $g_0 + \frac{t_d}{\epsilon_r}$ (Fig. 3.1b) and it is independent of any other system parameters.

The pull-in voltage is one of the most important parameters of RFMEMS design. As Eq. 3.3 indicates, it depends on the initial plates separation $g_0$, the spring constant $k$ and the actuation area $a$. Since $g_0$ is usually fixed by the sacrificial layer thickness, the only free parameters are $k$ and $a$. Typically high values of $k$ are used, since they provide large restoring forces necessary to overcome possible stiction due to dielectric charging or moisture, as mentioned in Section 3.1. Consequently, the actuation area $a$ of RFMEMS must be increased to prevent a very high pull-in voltage due to a large spring constant. Thus, there is a trade-off between compactness, low operation voltage and high restoring force. As a result, pull-in voltages of most RFMEMS are between 20..100 V.

### 3.2.2 Dynamic Pull-in

As mentioned above, static analysis assumes that the applied voltage is increased slowly and inertia and damping have no effect on the movement of the membrane. However, in practice electrostatic MEMS are actuated by a voltage step-function, causing fast movement of the membrane. Thus inertia and damping must be taken into account. If damping is small and inertia is large, it is expected that the membrane would overshoot the static equilibrium position, at least for a short time. Besides, in case of low damping the pull-in voltage should be reduced due to the inertial force acting in the same direction as the electrostatic force. These intuitive expectations are theoretically confirmed below.

The following analysis of the dynamic response for the case of a step-function actuation mainly follows [37–39]. The total energy of the mass-spring system in Fig. 3.1, i.e. the Hamiltonian $H$, is a sum of the kinetic energy, the potential energy of the spring and the electrostatic potential energy of the MEMS capacitor (as in the static analysis, fringing electric fields are neglected for brevity):

$$H = \frac{1}{2}m\left(\frac{\mathrm{d}x}{\mathrm{d}t}\right)^2 + \frac{1}{2}kx^2 - \frac{1}{2}\frac{\epsilon_0 a V^2}{\left(g_0 + \frac{t_d}{\epsilon_r} - x\right)} \quad (3.4)$$

If damping is small and can be neglected, the Hamiltonian of the system does not change with time. This assumption is often correct, since MEMS are operated mostly in air at atmospheric pressure and in addition, their moveable membranes have small holes for etching of the sacrificial layer, that significantly reduces the air squeeze-film damping. The assumption of small damping was also validated in our experiments presented in Chapters 4 and 5.

Thus, the value of the Hamiltonian at time $t = 0$ (when the voltage step-function is applied) can be used as the energy constraint of the system. Its value is easily found from Eq. 3.4 by setting deflection $x_{|t=0}$ and velocity $\frac{\mathrm{d}x}{\mathrm{d}t}_{|t=0}$ to zero:

$$H(t_0) = -\frac{1}{2}\frac{\epsilon_0 a V^2}{\left(g_0 + \frac{t_d}{\epsilon_r}\right)} \quad (3.5)$$

Substituting Eq. 3.5 for $H$ into Eq. 3.4 results in the dynamic response function:

$$m\left(\frac{\mathrm{d}x}{\mathrm{d}t}\right)^2 + kx^2 - \frac{\epsilon_0 a V^2 x}{\left(g_0 + \frac{t_d}{\epsilon_r} - x\right)\left(g_0 + \frac{t_d}{\epsilon_r}\right)} = 0 \quad (3.6)$$

In [37–39], the solutions for $x(t)$, i.e. the trajectories of the moveable membrane, were calculated with MATLAB by numerical time-integration of the acceleration $\frac{\mathrm{d}^2 x}{\mathrm{d}t^2}$. The acceleration

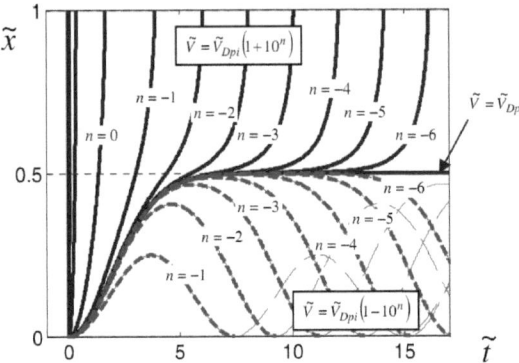

Figure 3.3: After [37–39]: Normalized deflection $\tilde{x}$ of the parallel-plates actuator shown in Fig. 3.1 versus normalized time $\tilde{t}$ for various values of applied voltage $\tilde{V}$, where $\tilde{V}_{Dpi}$ is the normalized dynamic pull-in voltage, $\tilde{x} = 0$ at the initial position and $\tilde{x} = 1$ at the bottom electrode. The details of normalization can be found in [37–39] and are not repeated here.

can be obtained by integrating Eq. 3.4 over time (which results in Newton's second law):

$$\frac{\mathrm{d}^2 x}{\mathrm{d}t^2} = -\frac{k}{m}x + \frac{1}{2}\frac{\epsilon_0 a V^2}{m\left(g_0 + \frac{t_d}{\epsilon_r} - x\right)^2} \tag{3.7}$$

The plot of exact trajectories can be found in [37–39] and is repeated here for completeness (Fig. 3.3). For voltages below the dynamic pull-in voltage, the moveable membrane oscillates around a certain deflection dependent on the voltage amplitude. Due to damping, these oscillations will eventually die out and the membrane will stay still at an equilibrium position. For voltages above the dynamic pull-in, the membrane collapses onto the bottom electrode. In the boundary case, i.e. at dynamic pull-in, the membrane goes to the equilibrium position without oscillations.

The stagnation deflection $x_s$ for an applied voltage $V$ can be found from Eq. 3.4 by setting the velocity $\frac{\mathrm{d}x}{\mathrm{d}t}$ to zero and solving for $x_s$:

$$x_s = \frac{k\left(g_0 + \frac{t_d}{\epsilon_r}\right)^2 - \sqrt{k^2\left(g_0 + \frac{t_d}{\epsilon_r}\right)^4 - 4k\left(g_0 + \frac{t_d}{\epsilon_r}\right)\epsilon_0 a V^2}}{2k\left(g_0 + \frac{t_d}{\epsilon_r}\right)} \tag{3.8}$$

The maximum of this function corresponds to the dynamic pull-in situation. Equating the expression under the square-root to zero, the dynamic pull-in voltage and corresponding deflection are found to be:

$$\begin{cases} x_{dyn.pull-in} = \dfrac{\left(g_0 + \frac{t_d}{\epsilon_r}\right)}{2} \\ V_{dyn.pull-in} = \sqrt{\dfrac{k\left(g_0 + \frac{t_d}{\epsilon_r}\right)^3}{4\epsilon_0 a}} \end{cases} \quad (3.9)$$

Comparing Eq. 3.9 to Eq. 3.3 reveals that the dynamic pull-in voltage is about 8% smaller than that derived from the static analysis. The maximum deflection corresponds to one-half of the initial plate separations, whereas for the static case it is only one-third. This confirms intuitive expectations that due to inertia the membrane can overshoot the static equilibrium position and the voltage required for the pull-in is smaller than the one obtained from the static analysis.

### 3.2.3 Switching Time

Switching time is an important parameter, critical especially in such RFMEMS applications as T/R switches or phase shifters. To derive the equation for the switching time, Eq. 3.6 can be re-written:

$$\frac{dx}{dt} = \sqrt{-\frac{k}{m}x^2 + \frac{\epsilon_0 a V^2 x}{m\left(g_0 + \frac{t_d}{\epsilon_r} - x\right)\left(g_0 + \frac{t_d}{\epsilon_r}\right)}} \quad (3.10)$$

It is common in the literature to define $\omega_0$ as the mechanical resonance frequency of the system in the undamped case:

$$\frac{k}{m} = \omega_0^2 \quad (3.11)$$

Then the switching time can be obtained by integration from the initial position to the final stagnation position $x_s$:

$$t_s = \int_0^{x_s} \frac{dx}{\omega_0 \sqrt{\frac{\epsilon_0 a V^2 x}{k\left(g_0 + \frac{t_d}{\epsilon_r}\right)\left(g_0 + \frac{t_d}{\epsilon_r} - x\right)} - x^2}} \quad (3.12)$$

If the applied voltage is above $V_{dyn.pull-in}$, then $x_s$ equals $g_0$. Otherwise $x_s$ is below $x_{dyn.pull-in}$ and is given by Eq. 3.8. In [39], the integration has been solved analytically:

$$t_s = \begin{cases} \frac{\pi+3\ln(2)-\ln(\delta_1)+O(\delta_1)}{2\omega_0}, & \text{for } V = V_{dyn.pull-in}(1-\delta_1) \\ \frac{3\ln(2)-\ln(\delta_2)+O(\delta_2)}{\omega_0}, & \text{for } V = V_{dyn.pull-in}(1+\delta_2) \end{cases} \quad (3.13)$$

where $\delta_1$ and $\delta_2$ determine $V$ as a function of $V_{dyn.pull-in}$ with $0 \leqslant \delta_1 < 1$ and $0 \leqslant \delta_2 \ll 1$. In addition, [39] gives numerical solutions to the integration, resulting in compact and accurate closed-form equations for the switching time:

$$t_s = \begin{cases} \frac{5.223-\ln(\delta_1)}{2\omega_0}, & \text{for } V = V_{dyn.pull-in}(1-\delta_1) \\ \frac{2.068-\ln(\delta_2)}{\omega_0}, & \text{for } V = V_{dyn.pull-in}(1+\delta_2) \end{cases} \quad (3.14)$$

### 3.2.4 Release Time

The release process can be described by free oscillations of the mass-spring system shown in Fig. 3.1. The governing equation is given below, where $\eta$ is the damping coefficient and $\omega_0 = \sqrt{\frac{k}{m}}$ is the resonance frequency of the system in the undamped case:

$$\frac{dx^2}{dt^2} + \eta\frac{dx}{dt} + \omega_0^2 x = 0$$

The solution is a harmonic function given by Eq. 3.15. Its angular frequency is $\omega_d$ and the amplitude is decaying with time according to $e^{-\frac{\eta}{2}t}$ term. The parameters $\omega_d$ and $\eta$ fully determine the release process of MEMS devices. When the DC-voltage is removed, the restoring spring force moves the plate upwards. The higher the frequency $\omega_d$, the faster is the release of the RFMEMS component. The higher the damping coefficient $\eta$, the faster is the settling of oscillations following the release. The constants $c_1$ and $c_2$ are found by satisfying the initial conditions for position and velocity. Typically, at $t = 0$ the moveable plate is at the down position and is not moving, i.e. $x_{|t=0} = g_0$ and $\frac{dx}{dt}_{|t=0} = 0$.

$$\begin{cases} x(t) = e^{-\frac{\eta}{2}t}\left(c_1\,cos(\omega_d t) + c_2\,sin(\omega_d t)\right) \\ \omega_d = \sqrt{\omega_0^2 - \dfrac{\eta^2}{4}} \\ c_1 = x_{|t=0} \\ c_2 = \dfrac{1}{\omega_d}\left(\dfrac{\eta}{2} x_{|t=0} + \dfrac{dx}{dt}_{|t=0}\right) \end{cases} \quad (3.15)$$

However, Eq. 3.15 assumes no variation of the damping coefficient $\eta$ with deflection. Generally, it is not true, since $\eta$ increases nonlinearly with increasing deflection. Still, the assumption of a constant damping for brevity is a common practice (e.g. [40]) and is also used in this work.

Switching and release dynamics of the developed RFMEMS components were studied experimentally using a Laser Doppler Vibrometer available at IHP Microelectronics [41]. Chapters 4 and 5 contain detailed analysis of the measured data, discuss switching and release time calculations and compare them with the presented theory.

## 3.3 Electromagnetic Analysis of Electrostatic MEMS

Electromagnetic analysis of RFMEMS has been covered in detail by many authors and only a brief summary is presented here. The overview below partly follows [6].

### 3.3.1 Shunt Capacitive Switches and Switchable Capacitors

Shunt capacitive switches and switchable capacitors are often realized as moveable air-bridges above a coplanar line, as schematically shown in Fig. 3.4. When no voltage is applied, the air-bridge, i.e. the moveable membrane, is in the up-state. If the applied voltage is above the dynamic pull-in voltage given by Eq. 3.9, the membrane snaps down onto the underlying electrode covered with a dielectric layer. Thus, a switchable capacitance can be obtained. When used as a switch, this configuration works best at higher (upper microwave and millimeter-wave) frequencies, where the impedance of the down-state capacitance is low enough to provide a good short-circuit necessary for high isolation.

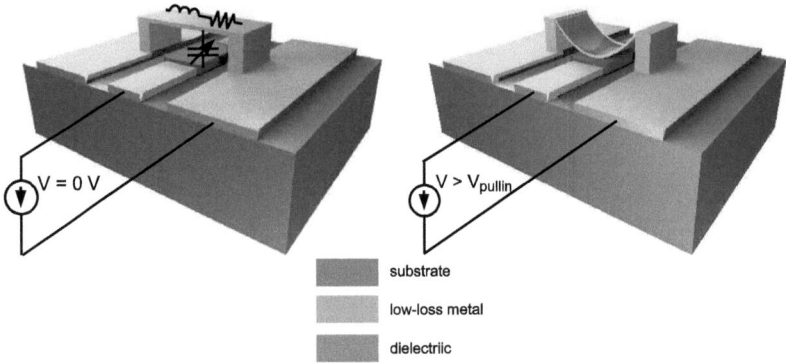

Figure 3.4: Schematic illustration of a MEMS bridge in shunt to a coplanar line in the up- (left) and down-states (right).

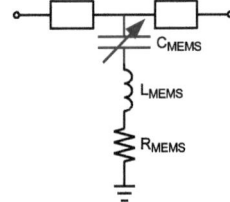

Figure 3.5: Equivalent circuit of a shunt capacitive MEMS switch or variable capacitor

A simplified equivalent circuit is shown in Fig. 3.5. The impedance of the shunt bridge is generally defined by Eq. 3.16 and its variation with frequency is qualitatively described by Eq. 3.17.

$$Z_{shunt} = \frac{1}{j\omega C} + j\omega L + R \qquad (3.16)$$

$$Z_{shunt} = \begin{cases} \frac{1}{j\omega C} & f \ll f_{res} \\ R & f = f_{res} \\ j\omega L & f \gg f_{res} \end{cases} \qquad (3.17)$$

where $f_{res}$ is the MEMS $LC$ resonance frequency:

$$f_{res} = \frac{1}{2\pi\sqrt{LC}} \qquad (3.18)$$

The up-state capacitance is typically small, which results in a very low insertion loss in the up-state. For a switch operation, the down-state capacitance typically should be 15-20 times larger than the up-state capacitance. In addition, a series LC resonance with the MEMS inductance leads to very good switch isolation in the down state, especially around the resonance frequency $f_{res}$. For a switchable capacitor operation, down-state capacitance is only 3-6 times larger than the up-state capacitance.

Values of the individual equivalent circuit elements can be obtained from S-parameters. In the up-state the impedance is mainly determined by the small shunt capacitance $C_{up}$. Hence, the corresponding reflection coefficient $S_{11}$ can be written as follows [42]:

$$S_{11,\,up} = \frac{-j\omega C_{up} Z_0}{2 + j\omega C_{up} Z_0} \qquad (3.19)$$

For $\omega C_{up} Z_0 \ll 2$, or equivalently for $|S_{11}| \leq -10$ dB, Eq. 3.19 can be simplified to [6]:

$$|S_{11,\,up}|^2 \approx \frac{(\omega C_{up} Z_0)^2}{4} \qquad (3.20)$$

Thus, using Eq. 3.20 the up-state capacitance can be easily calculated from measured or simulated S-parameters.

The down-state capacitance can be extracted from the down-state S-parameters. For frequencies well below the resonance $f_{res}$, the transmission coefficient $S_{21}$ is mainly determined by the shunt capacitance $C_{down}$ [42]:

$$S_{21,\,down} = \frac{2}{2 + j\omega C_{down} Z_0} \qquad (3.21)$$

For $\omega C_{down} Z_0 \gg 2$, or equivalently for $|S_{21}| \leq -10$ dB, Eq. 3.21 simplifies to [6]:

$$|S_{21,\,down}|^2 \approx \frac{4}{(\omega C_{down} Z_0)^2} \qquad (3.22)$$

Thus, using Eq. 3.22 the down-state capacitance can be directly calculated from measured or simulated S-parameters.

Alternatively, up- and down-state capacitances can be obtained from Z-parameters by considering the MEMS bridge as a generic T-network. Then impedance of the shunt branch is given by the $Z_{21}$ parameter and the MEMS capacitances can be found as:

$$\begin{cases} C_{up} = -\dfrac{1}{imag(Z_{21,up})\omega} & f \ll f_{res} \\ C_{down} = -\dfrac{1}{imag(Z_{21,down})\omega} & f \ll f_{res} \end{cases} \quad (3.23)$$

It should be noted that the extracted capacitances also contain a contribution from the distributed transmission line capacitance $C'l$ where $l$ is the length of the transmission line below the MEMS bridge. Thus, for an accurate estimation of the MEMS capacitance, $C'l$ should be calculated as shown in Section 8.1.1 and subtracted from the values obtained with Eq. 3.20, Eq. 3.22 or Eq. 3.23.

The MEMS inductance can be calculated from the measured or simulated resonance frequency (Eq. 3.18) and the down-state capacitance. The series resistance can be obtained from the down-state transmission coefficient $S_{21}$ at resonance, assuming $R \ll Z_0/2$:

$$S_{21\ down,\ res} = \frac{2}{2 + Z_0/R} \approx \frac{2R}{Z_0} \quad (3.24)$$

### 3.3.2 Series Switches with Metal-to-Metal Contact

Series switches with metal-to-metal contact are often constructed as a coplanar line with an interrupted centre conductor and a MEMS cantilever, as schematically illustrated in Fig. 3.6. When an actuation voltage is applied, the cantilever snaps down and closes the gap in the coplanar line through a metal-to-metal contact. Such switch works best up to microwave frequencies, since isolation considerably degrades as frequency increases to millimeter-waves due to an increased series capacitive coupling.

A simplified equivalent circuit of a metal-to-metal contact MEMS cantilever is shown in Fig. 3.7. $Z_{high}$ models the cantilever as a transmission line of high impedance, $C_{up}$ is the up-state capacitance due to the overall of the cantilever's tip with the contact region, $C_{coupling}$ accounts for capacitive coupling through substrate and $R_{down}$ is the contact resistance in the down-state. The impedance in the up-state state is dominated by the small capacitance $C_{up}$.

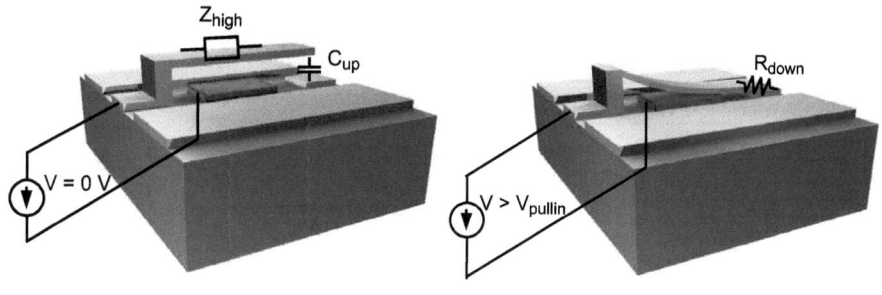

Figure 3.6: Schematic illustration a MEMS cantilever in series with a coplanar line in the up- (left) and down-states (right).

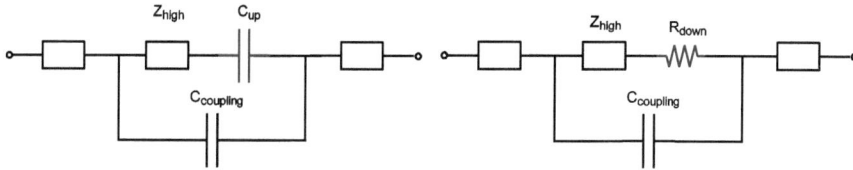

Figure 3.7: Equivalent circuit of a series MEMS switch with a metal-to-metal contact in the up-(left) and down states (right).

Thus, the transmission coefficient $S_{21}$ can be calculated as follows [42]:

$$S_{21,\,up} = \frac{j2\omega C_{up} Z_0}{1 + j2\omega C_{up} Z_0} \qquad (3.25)$$

For $2\omega C_{up} Z_0 \ll 1$, or equivalently for $|S_{21,\,up}| \ll -10\ dB$, the reflection coefficient can be approximated as [6]:

$$|S_{21,\,up}|^2 \approx (2\omega C_{up} Z_0)^2 \qquad (3.26)$$

allowing to calculate the up-state capacitance from the measured or simulated isolation.

The down-state resistance $R_{down}$, which includes contact resistance and the losses in the membrane, can be calculated from the down-state loss of the switch given by [6]:

$$Loss_{down} = 1 - |S_{21,\ down}|^2 - |S_{11,\ down}|^2 \approx \frac{R_{down}}{Z_0} \qquad (3.27)$$

# Chapter 4

# Components in ISiT RFMEMS Technology

This chapter presents components developed using RFMEMS technology of the Fraunhofer Institute for Silicon Technology (ISiT), Itzehoe, Germany, carried out in cooperation during the European FP6 projects AMICOM and RF-Platform. The ISiT technology is undergoing a constant development and is not yet commercially available. This work contributed to technology improvement by assessing microwave losses due to a conductive channel at Si-SiO$_2$ interface. Besides, the portfolio of ISiT RFMEMS components was extended with switches with optimized biasing scheme, variable capacitors and a series cantilever.

## 4.1 ISiT RFMEMS Fabrication Process

### 4.1.1 ISiT RFMEMS Fabrication Steps

The RF-MEMS technology of the Fraunhofer Institute for Silicon Technology (ISiT), Itzehoe, Germany is based on surface micromachining [43]. The substrate in the ISiT process is an oxidized high-resistivity silicon (> 3 k$\Omega$·cm, thickness 508 μm). As described in Section 4.1.3, treatment of high-resistivity Si substrates with either poly-Si or Argon is the first processing step. After that high-resistivity ($50$ $\Omega$/sq) tantalum is deposited to be used for MEMS bias lines. The subsequent steps of RFMEMS fabrication are schematically illustrated in Fig. 4.1 and also described below:

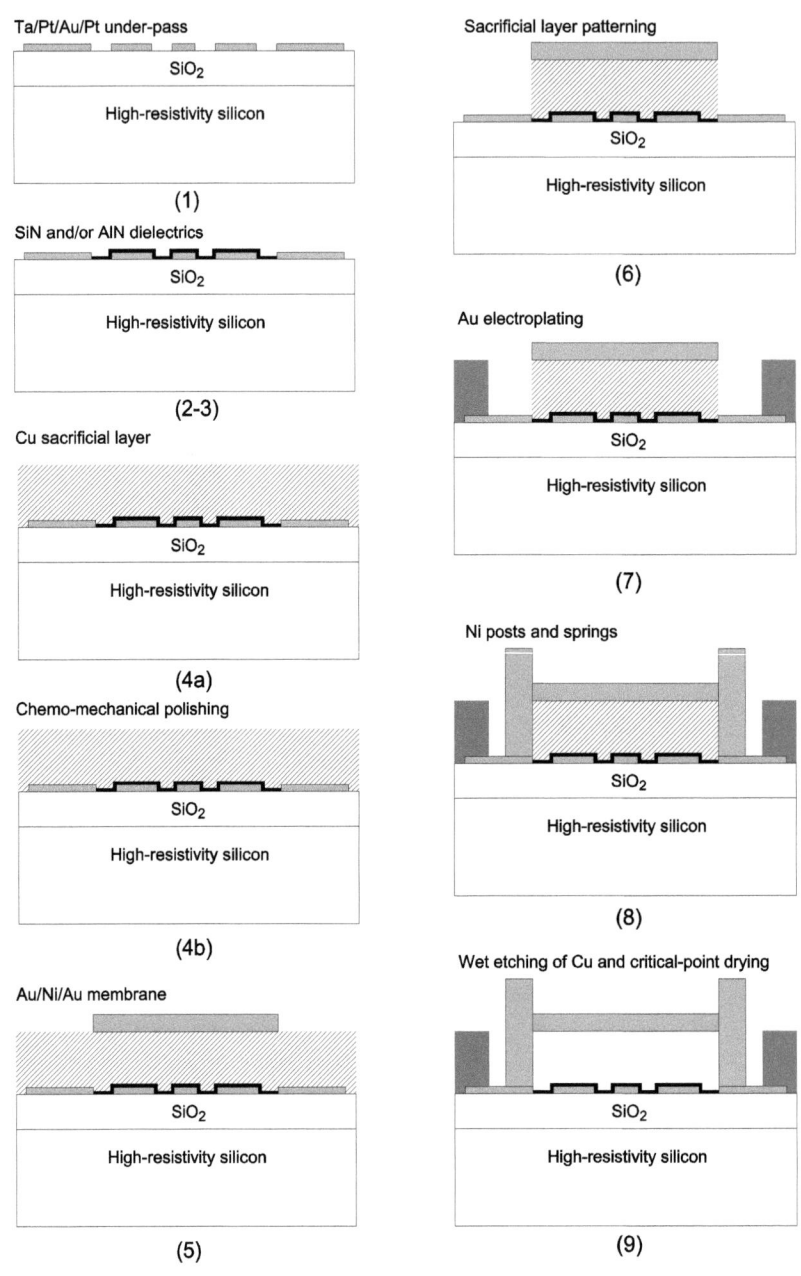

Figure 4.1: Fabrication steps of the Fraunhofer ISiT RFMEMS process (not to scale)

1. For RFMEMS under-pass a stack of Ta/Pt/Au/Pt (thickness 20 nm / 40 nm / 300 nm / 60 nm) is patterned by lift-off.

2. PECVD SiN (300 μm, $\epsilon_r = 7.4$) is deposited next.

3. The second dielectric layer is AlN (300 μm, $\epsilon_r = 10$). Compared to SiN, AlN is much less prone to dielectric charging [44] and is used for DC-isolation of actuation electrodes. Due to a larger dielectric constant, it is also used to form the MEMS capacitance.

4. Copper is used as the sacrificial layer. First, 3 μm-thick Cu is deposited and then chemical mechanical polishing (CMP) is used to ensure a smooth surface for the proceeding membrane deposition.

5. The RFMEMS moveable membrane made of a Au/Ni/Au stack (thickness 200 nm / 500 nm / 200 nm respectively) with 0.06 $\Omega$/sq sheet resistance and 500 MPa tensile stress is evaporated.

6. The sacrificial layer is patterned by wet etching.

7. The signal line outside the switching area and all other interconnections of 2 μm thick Au are electroplated.

8. Springs, anchoring and free-standing parts of the switch are fabricated of 15 μm thick Ni, which has nearly zero stress and very high stiffness.

9. Finally, the sacrificial copper layer is removed by wet-etching followed by a critical-point drying, necessary to avoid membrane sticking due to adhesion forces.

## 4.1.2 ISiT Standard Switch

Over several years ISiT has been working on improving the RFMEMS process by investigating different MEMS structural materials and fabrication parameters. It was found that the commonly used dielectric SiN is prone to charging, leading to a residual electric field in thin dielectric layers and thus, low MEMS life time [45]. Thus, SiN was replaced with AlN, which has a much better performance in terms of long-time reliability [44]. As structural material for moveable membranes, a stack of gold/nickel/gold was designed to reduce ohmic losses due to the high conductivity of gold and to ensure good mechanical properties due to the high Young's modulus of nickel. Besides, temperature compensation techniques for springs and suspended membranes were developed, which minimize stresses in the membrane after the fabrication and packaging and ensuring good temperature stability of MEMS structures [46].

As a result of the extensive work, a MEMS shunt switch has been developed (Fig. 4.2). The measured microwave performance of the switch is shown in Fig. 4.3. The insertion loss is below 0.8 dB up to 50 GHz and maximum isolation of about 33 dB is achieved at about 42.5 GHz. This switch has been optimized by ISiT for 40..50 GHz operation. However, other important bands were not yet covered at the outset of this work. Also, switchable capacitors needed for loaded-line phase shifters and impedance tuners have not yet been developed. So, this switch provided a basis for the subsequent designs. Most importantly, the special layout of thermally-compensated membrane supporting springs was used in this work.

In addition, the existing switch had a complicated biasing scheme, requiring three DC-signals: 65 V applied to two separate actuation electrodes and 20 V applied to the signal line. The advantage of this configuration is decoupling of the RF signal from the actuation area, which greatly increases switch robustness against self-actuation and RF hold-down (latching) as compared to the case when the transmission line is used as an actuation electrode. In the former case the membrane can be much stiffer for the same pull-in voltage, resulting in higher restoring force. However, the biasing scheme of the ISiT switch is less suitable for circuits employing several independent switches along one signal line, since it would require a lot of DC-blocks, increasing losses of the RF-signal.

Thus, one of the goals of this work work to extend the ISiT-MEMS portfolio with switches for other frequency bands and also with switchable capacitors. Besides, the actuation scheme had to be simplified to preferably one DC-signal per MEMS component and to avoid high DC-voltage at the RF signal line.

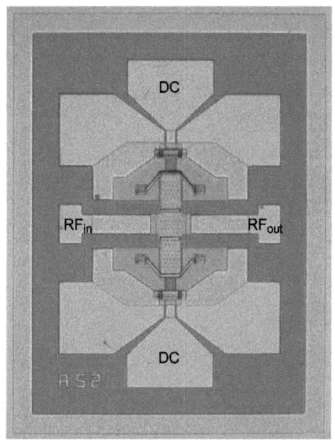

Figure 4.2: Micrograph of a shunt capacitive switch developed at ISiT

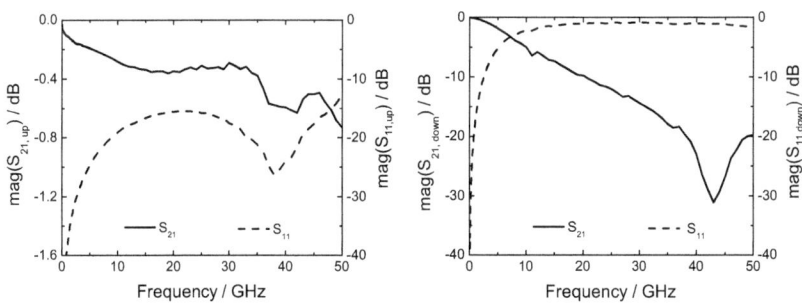

Figure 4.3: Measured S-parameters of the standard ISiT switch shown in Fig. 4.2 in the up- (left) and down- (right) states

### 4.1.3 Microwave Losses of Oxidized High-resistivity Silicon

Our initial technology assessment revealed large substrate losses due to formation of a conducting channel at the Si-SiO$_2$ interface. Several solutions to reduce the channel conductivity have been proposed in the literature, e.g. [47,48]. ISiT adopted two ways of silicon surface treatment before oxide deposition:

- LPCVD of a 500 nm thick, nominally undoped poly-crystalline silicon layer (poly-Si)
- implantation of Argon ($10^{15}$ cm$^{-2}$, 80 keV)

Fig. 4.4 compares attenuation constants of a test coplanar line measured on substrates with different treatments and simulated using a built-in Agilent ADS model for a lossless silicon substrate. It indicates that both surface treatment techniques minimized losses due to the conducting channel and restored the low-loss performance of high-resistivity silicon. Detailed results of experimental characterization up to 110 GHz were reported by our group in [49].

However, deposition of either poly-Si or implantation of Argon lead to increased surface roughness. This sightly increases resistive losses of the metal line. However, it follows from Fig. 4.4 that reduction of dielectric losses fully compensates the line resistance increase, since the plotted attenuation constants include contributions from all loss mechanisms (mainly dielectric, resistive and radiation). More important is the fact, that higher surface roughness results in a lower down-state capacitance and thus, lower capacitance ratio between up- and down-states. Still, since low loss performance has higher priority, additional treatment of silicon surface with either poly-Si or Argon has become a standard fabrication step in the ISiT process.

## 4.2 Shunt Capacitive SPST Switch

### 4.2.1 Design

This section presents the design of a shunt capacitive switch with a simplified biasing scheme. SEM-images in Fig. 4.5 show the fabricated MEMS air-bridge above a coplanar

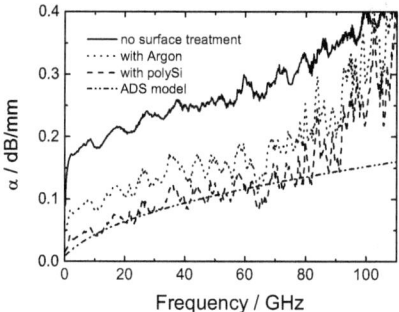

Figure 4.4: Measured CPW attenuation constants for w=88 μm, s=44 μm (w: centre conductor width, s: slot width) for different substrate treatments and comparison to ADS simulation of a built-in CPW model for a lossless Si substrate.

line. The membrane has small holes, necessary for the sacrificial layer etching below the membrane and also to reduce squeeze film damping of air for small switching time. Thick Ni stiffening bars were used to prevent membrane's lateral bowing typical for thin membranes with tensile stress. Major modifications of the moveable membrane can alter built-in stresses in the thin metal layer, leading to different stiffness and even lateral bowing and buckling in the worst case. Thus, the shape of the moveable membrane and its supporting springs were adopted from the standard ISiT switch. However, RF- and DC-configurations of the standard switch were redesigned to keep only one DC actuation signal.

First modifications of the DC-bias network are described. Depending on the biasing scheme the MEMS bridge can either be grounded, i.e. connected to coplanar line grounds, or floating. In case of grounded membranes, the actuation voltage is applied either to the RF signal line located below the membrane or to two separate actuation electrodes, as in case of the ISiT standard switch. In the former case, a circuit employing several switches which must be controlled individually would require DC-blocking capacitors between the switches. Besides, high DC-actuation voltage on the RF-signal line is not desirable. In case of separate actuation electrodes, two bias lines per switch are required, all of that making the design more cumbersome.

An alternative solution used here is to employ floating membranes, where the actuation signal is applied to the MEMS bridge, while the RF signal line is kept at DC-ground. This is

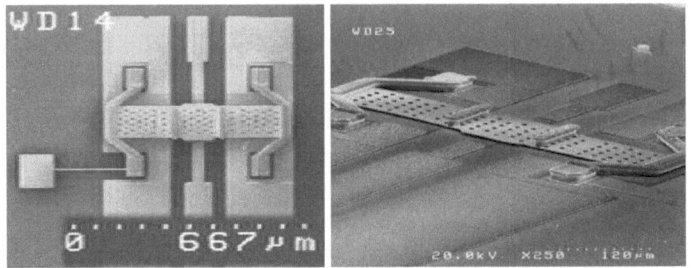

Figure 4.5: SEM images of the fabricated shunt capacitive SPST switch from Section 4.2

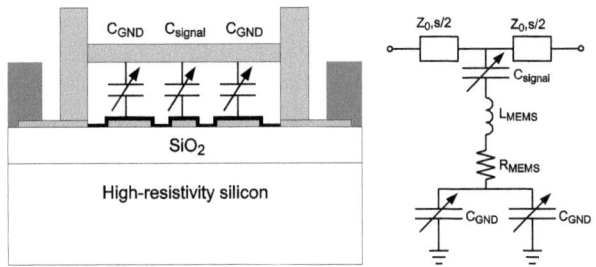

Figure 4.6: Cross-section (left) and equivalent circuit (right) of the shunt capacitive MEMS switch shown in Fig. 4.5

illustrated in the schematic cross-section of the switch shown in Fig. 4.6 (left). The shades of gray in the cross-sectional drawing correspond to the same materials as in Fig. 4.1. As Fig. 4.6 (left) shows, the MEMS bridge is DC-isolated from the coplanar line. The coplanar line is covered by an AlN dielectric layer and the moveable membrane is made of the Au/Ni/Au stack. As compared to the ISiT standard switch, no separate actuation electrodes are used and the DC-bias is applied directly to the membrane. The coplanar line below the bridge is kept at DC-ground. As a result, the actuation area is formed by the two grounds and the center conductor of the coplanar line. Thus, the actuation area is larger as compared to the standard ISiT switch, having smaller sized actuation electrodes only below the outer part of the membrane. This decreases the actuation voltage without reducing the membrane's spring constant, thus increasing MEMS robustness against dielectric charging. Another advantage of such bias arrangement is that only one DC-signal is needed and that high DC-voltage is not present at the RF signal line, which significantly simplifies subsequent system integration.

The equivalent circuit of this switch is shown in Fig. 4.6 (right). The insertion loss and isolation of the switch are mainly determined by the capacitances in the up- and down-states respectively. The membrane is RF-grounded through a series combination of the centre-line capacitance and two CPW ground-plane capacitances. Thus, the overall capacitance is reduced as compared to the ISiT standard switch, which has a grounded membrane. As for the down-state, the reduced capacitance results in an increased resonance frequency of the MEMS bridge (Eq. 3.18) and thus the region of maximum isolation moves towards higher frequencies. In the up-state a reduced capacitance is always beneficial, since it reduces reflection losses, especially at high frequencies.

To obtain a good 50 $\Omega$ match in the up-state, short sections of transmission lines with a high characteristic impedance $Z_0$ are employed. The lines are needed for the following reasons. On one hand, the up-state capacitance should be as small as possible for low reflection losses in the up-state. On the other hand, the down-state capacitance should be high to obtain high isolation. Thus, the technologically-fixed capacitance ratio and required isolation in the down-state constrain the up-state capacitance choice. The solution is to use a slightly higher up-state capacitance and to compensate it with series inductors, which are realized as transmission lines of high characteristic impedance. The switch design was carefully

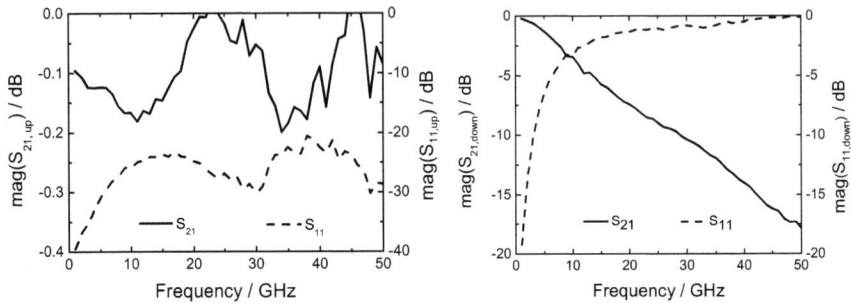

Figure 4.7: Measured S-parameters of the shunt switch shown in Fig. 4.5 in the up- (left) and down- (right) states. Fluctuations of $S_{21}$ are within measurement accuracy.

optimised with EM simulations to find a compromise between compactness (length) and losses (characteristic impedance, i.e. line width), of the inductive lines.

### 4.2.2 Electromagnetic Characterisation

The switch was characterized up to 50 GHz with the Agilent 8510C vector network analyser using SOLT (short-open-load-thru) calibration up to the probe tips. The measurement setup consisted of two GSG probes for the RF-signal and one DC-needle for the bias voltage.

The dynamic pull-down voltage of the switch is about 40 V. To ensure a good contact force in the down-state, a slightly higher value of 44 V is used for MEMS actuation. Measured S-parameters are shown in Fig. 4.7. It should be mentioned that sometimes unphysical positive "dB"-values of $S_{21}$ are within measurement accuracy.

In the up-state, the insertion loss is below 0.2 dB and return loss better than -20 dB up to 50 GHz. This corresponds to an up-state capacitance $C_{up}$ of 45 fF (Eq. 3.23). In the down-state, the isolation is 13 dB at 35 GHz and reaches 18 dB at 50 GHz, corresponding to a down-state capacitance $C_{down}$ of 720 fF (Eq. 3.23). The inductance and resistance of the MEMS have to be extracted from the resonance frequency, which is above the measurement limit. The measured down-state isolation is lower than was expected owing to increased surface roughness after the poly-Si deposition, which reduces the down-state capacitance,

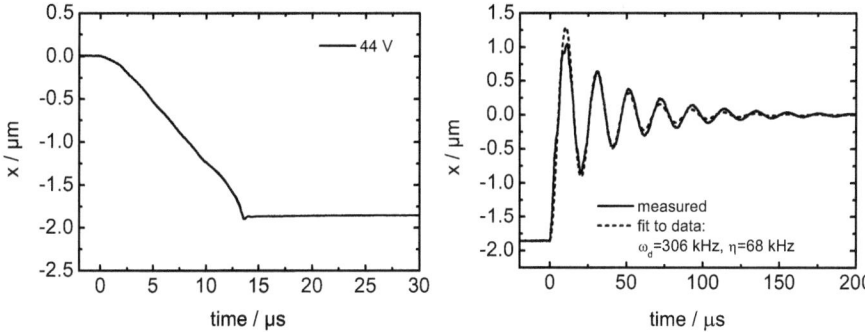

Figure 4.8: LDV measurements of switching (right) and release (left) of the shunt capacitive switch shown in Fig. 4.5

as described in Section 4.1.3. On the other hand, a very low insertion loss indicates that the losses due to a conducting channel at the $Si/SiO_2$ interface were eliminated.

### 4.2.3 Electromechanical Characterisation

Dynamic trajectories of the membrane movement during switching and release were measured with a Laser - Doppler - Vibrometer (LDV) available at IHP Microelectronics. During the experiment, the laser beam was focused at the center of the membrane. Rectangular pulses of 44 V with rising-/falling edges below 1 μs were applied to the actuation electrode to pull down and release the switch respectively.

The measured membrane deflections in the pull-down and release processes are shown in Fig. 4.8. The time axis is re-scaled so that the origin corresponds to either rising or falling edges of the applied rectangular pulse. Since the applied voltage is above the dynamic pull-in, the membrane goes to the down state without oscillations, in consistence with theoretical trajectories presented in Fig. 3.3. The measured switching time is about 13 μs. During the release time, the membrane undergoes free damped oscillations. Due to low squeeze-film damping of air, the membrane has an under-damped response. Thus, it takes several oscillations for it to settle at the up-state position. Consequently, the settling time is about 150 μs and is significantly larger than the switching time. However, relevant for practical applications is the release time, i.e. the time when the RF power starts to flow, which depends on the RF signal frequency, and is not identical to the mechanical settling time, as it is shown further down.

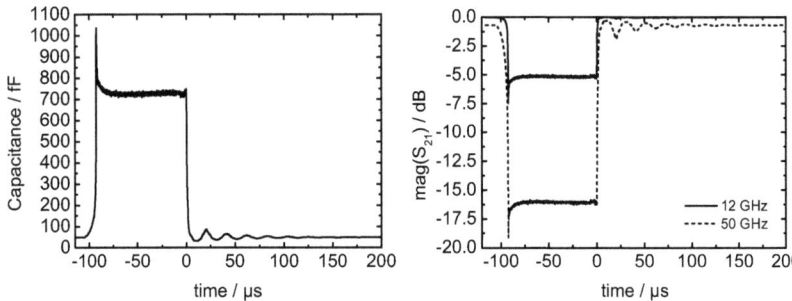

Figure 4.9: Calculated dynamic variations of capacitance (left) and transmission coefficient (right) during switching and release of the switch shown in Fig. 4.5

The release experimental data shown in Fig. 4.8 was fitted with Eq. 3.15 using nonlinear least squares. As explained in Section 3.2.4, Eq. 3.15 assumes a constant damping coefficient for all deflections that leads to small deviations between fitted and measured curves. The fitted damped oscillation frequency $\omega_d = 306$ kHz and the damping coefficient $\eta = 68$ kHz.

Due to typically low damping, the resulting mechanical oscillations of RFMEMS may lead to undesired modulation of the RF signal. The release time typically refers to the moment when the RF signal modulation settles down within acceptable amplitude variation. This time strongly depends on the frequency of the RF signal, as shown in the following.

Fig. 4.9 (left) shows the dynamic capacitance variation, calculated from the measured displacement using a parallel-plate formula and accounting for the down-state capacitance reduction due to surface roughness. As it can be seen, the largest capacitance variation takes place within about 50 μs after the release and it settles down afterwards. Fig. 4.9 (right) illustrates "dynamic" S-parameters calculated for 12 GHz and 50 GHz using Eq. 3.22 and the time-variant capacitance from Fig. 4.9 (left). At 12 GHz, there is little modulation in the transmission coefficient during release oscillations and the release time is as low as 3 μs. The same result was measured using a real-time oscilloscope for a 12 GHz RF signal, as shown in Fig. 4.10. At 50 GHz, the effect of capacitance modulation is much more pronounced and the release time is close to 50 μs. Still, both values are significantly below the mechanical settling time of 150 μs. Thus, the release time of RFMEMS can not be uniquely defined, since it strongly depends on the intended frequency of operation.

If Laser-Doppler-Vibrometer is not available, switching time can be estimated using closed-form equations developed by V. Leus et al. in [39] and discussed in Section 3.2.3. In the following, measured and theoretical switching times are compared to evaluate the accuracy of the equations reported in [39]. Switching time can be calculated using Eq. 3.14, which requires knowledge of the mechanical resonance frequency $\omega_0$, i.e. spring constant $k$ and mass $m$ of the MEMS component. The spring constant $k$ was found to be 170 N/m using Eq. 3.9 ($g_0$=2.5 μm, $t_d$=300 nm, $\epsilon_r$=10, $A$=48,600 μm²) and a dynamic pull-in voltage of 40 V. From the switch geometry and material density, the mass of the moving part $m$ was calculated to be 1.56 μg. Then the mechanical resonance frequency is 330 kHz. The mechanical resonance frequency $\omega_0$ can also be found using Eq 3.15 and experimental values of the damped oscillation frequency $\omega_d$ of 306 kHz and the damping coefficient $\eta$ of 68 kHz extracted from the release measurement shown in Fig. 4.8. The resulting value of $\omega_0$ is 308 kHz, which is quite close to 330 kHz, calculated from $k$ and $m$, confirming the validity of the approach.

Finally, for an actuation voltage of 44 V, i.e. for $\delta_2 = 0.13$, the predicted switching time $t_s$ is 12.6 μs. This value corresponds very accurately to the switching time measured using LDV and proves that an accurate estimation of the switching time from dynamic pull-in voltage and switch geometry is feasible. Thus, this simple method can be used, if LDV equipment is not available.

## 4.3 Shunt Capacitive SPST Switch for a 26 GHz Phase Shifter

### 4.3.1 Design

The design described below was optimized for the 26 GHz phase shifter presented in Section 9.2. The least significant bit of the phase shifter is set by the phase delay of one switch

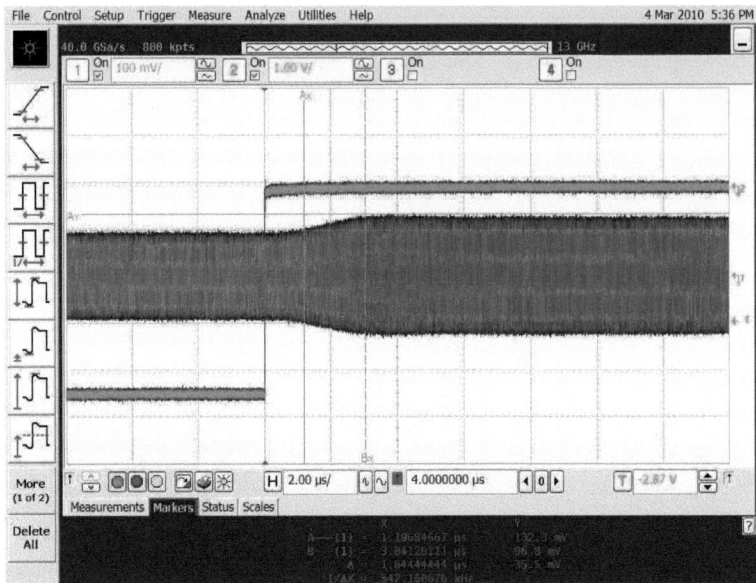

Figure 4.10: Release measurement of the switch shown in Fig. 4.5 using a real-time oscilloscope for a 12 GHz RF input signal. Actuation signal is shown in *green* and the RF signal is shown in *brown*. The time scale is 2 μs per division.

in the up-state, which should be as small as possible. Besides, a good isolation in the down-state is needed, but simulations showed that isolation of about 15 dB at 26 GHz is sufficient.

An SEM-image and a micrograph of the fabricated switch are shown in Fig. 4.11. The design is similar to the one presented in Section 4.2, but the membrane shape was optimized to achieve RF-performance according to the phase shifter requirements. AlN was used as the dielectric layer and the membrane is made of thin Au/Ni/Au stack. To benefit from the inherent DC-isolation of MEMS actuation signals from the RF-signal, a floating membrane is used.

Cross-section and equivalent circuit of the switch are identical to the ones shown in Fig. 4.6. The switched MEMS capacitance is formed by a series connection of $C_{signal}$ and two $C_{GND}$, which are the capacitances between the MEMS bridge and CPW centre conductor and grounds respectively. The series connection reduces the overall MEMS capacitance as

Figure 4.11: Micrograph and SEM image of the 26 GHz shunt capacitive switch

compared to a grounded-membrane, where only $C_{signal}$ is present. This is advantageous for the up-state capacitance, since a low $C_{up}$ is desired for better resolution of the phase shifter. However, a high capacitance is required to provide a short-circuit in the down-state at the operation frequency (26 GHz). The reduction of the overall capacitance due to two $C_{GND}$ is minimized by increasing the ground areas.

As explained inn Section 9.2, the resolution of the phase shifter is determined by the phase delay of a single switch in the up-state and the interconnecting transmission line. To increase phase shifter resolution, up-state capacitance $C_{up}$ and the spacing between neighbouring switches $d$ should be as small as possible. The spacing is set to 262 µm, which is the technological minimum for the supporting springs dimensions. So, the value of the up-state capacitance is the most critical parameter for the switch design. Since the ratio of up- and down-state capacitances is fixed, reduction of the up-state capacitance simultaneously lowers the down-state capacitance, which then may not provide enough isolation at 26 GHz. For narrow-band applications the reduction in the down-state capacitance can be partly compensated by increasing the switch inductance by narrowing the membrane above the coplanar waveguide gaps [50]. This lowers the LC resonance frequency of the switch and increases the isolation at lower frequencies. With the help of EM-simulations, an optimum MEMS membrane geometry with low up-state capacitance and high inductance was found, which can be seen in the micrograph and SEM-image in Fig. 4.11.

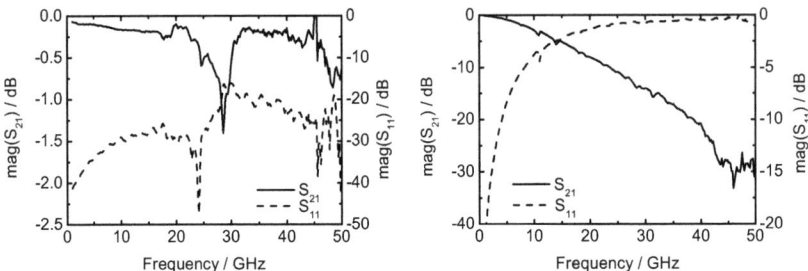

Figure 4.12: Measured S-parameters of the switch shown in Fig. 4.11 in the up- (left) and down- (right) states

### 4.3.2 Electromagnetic Characterisation

The switch was characterized up to 50 GHz with the Agilent 8510C vector network analyser using SOLT calibration up to the probe tips. The measurement setup consisted of two GSG probes for the RF-signal and one DC-needle to apply the actuation voltage.

The measured performance of the MEMS switch is shown in Fig. 4.12. In the up-state, the insertion loss is better than 0.2 dB up to 40 GHz, with the exception of a weak resonance at about 29 GHz, whose origin has not yet been found. The return loss is better than 16 dB up to 50 GHz. In the down-state, the isolation reaches 33 dB at 46 GHz, i.e. where the capacitance and inductance of the switch resonate. At 26 GHz, the isolation is about 15 dB, as required by the phase shifter design.

The extracted up- and down-state capacitances of the switch are 30 fF and 550 fF respectively (Eq. 3.23). The MEMS inductance is about 22 pH (Eq. 3.18). The series resistance is obtained from the transmission coefficient at the resonance frequency and is below 0.5 $\Omega$ (Eq. 3.24).

### 4.3.3 Electromechanical Characterisation

For this component Laser-Doppler-Vibrometer measurement was not available, thus a switching time was estimated using the theory described in Section 3.2.3. This theory was also confirmed experimentally in Section 4.2. The dynamic pull-in voltage of the switch is about 37 V and a 20% higher value of 45 V was used for the down-state measurement to increase the contact force for a better contact of the membrane with the dielectric layer and to increase the switching time. Using Eq. 3.9, the spring constant was extracted to be 170 N/m ($g_0$=2.5 µm, $t_d$=300 nm, $\epsilon_r$=10, $A$=60,920 µm$^2$). From the switch geometry and material density, the mass of the moving part $m$ is 1.73 µg. Then using Eq. 3.11 and Eq. 3.14, the mechanical resonance frequency $\omega_0$ is 325 kHz and the calculated switching time $t_s$ for 45 V ($\delta_2 = 0.22$) is 11.1 µs.

## 4.4 Tri-State Shunt Switchable Capacitor

### 4.4.1 Design

This switchable capacitor was designed for the DMTL impedance tuner presented in Section 10.3 and phase shifter presented in Section 9.3. A distinguishing feature of this component is tri-state operation. Depending on the actuation scheme, three distinct capacitances with a ratio of 1:4.5:16 can be obtained. Besides, the capacitances in two of the possible states almost do not depend on the internal membrane stress, which can differ across the wafer. This property is very important for high yield of RFMEMS components.

A schematic cross-section of the switchable capacitor is shown in Fig. 4.13. The CPW ground planes are covered with both SiN and AlN dielectric layers available in the ISiT process with the total thickness of 600 µm. The center conductor is covered only with AlN (thickness of 300 nm). Thus, if in the down-state the membrane is kept flat, the down-state MEMS capacitance will be significantly reduced due to the remaining air gap of 300 nm. To obtain such a situation, the center conductor of the coplanar line must be biased at the same potential as the MEMS bridge, indicated by $V$ as shown in Fig. 4.13 (b). On the other hand, when the center conductor is kept at DC-ground (Fig. 4.13 (c)), there is an attractive force, causing the membrane to bend down and touch the underlying AlN layer. Thus, the capacitance increases. As a consequence, two of the possible capacitances are set by the air-gap and are only slightly dependent on the membrane shape variations due to the non-uniform bowing.

SEM images of the fabricated RFMEMS switchable capacitor are shown in Fig. 4.14. The standard ISiT switch membrane having a well-controlled built-in stress was used. Stiffening

Ni-bars are added to prevent the membranes from lateral bowing. Short high impedance lines were used to compensate for the up-state capacitance and to obtain a good 50 Ω match in the up-state, when the capacitance is the smallest. In two out of three possible states, the DC-bias is applied to the membrane and only one DC-voltage per MEMS is needed. In the third capacitance state, a DC-bias is also applied to the RF-line, as explained below.

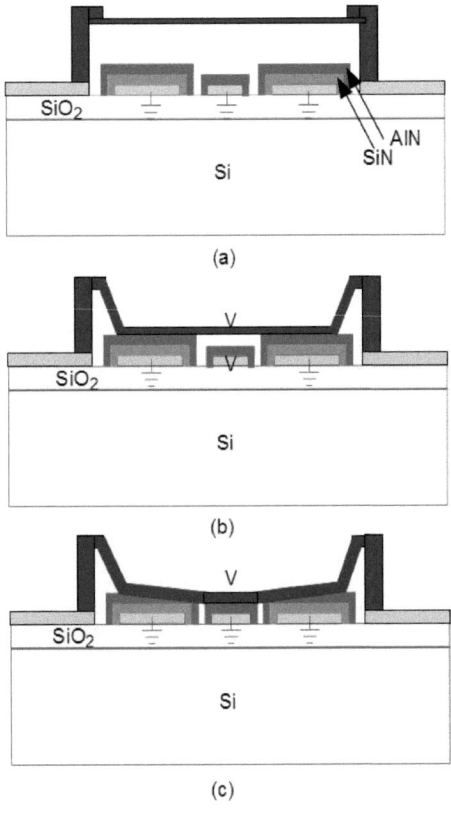

Figure 4.13: Schematic cross-section of the tri-state shunt capacitor. (a): without bias voltage, (b): with bridge and centre conductor biasing and (c): with only bridge biasing

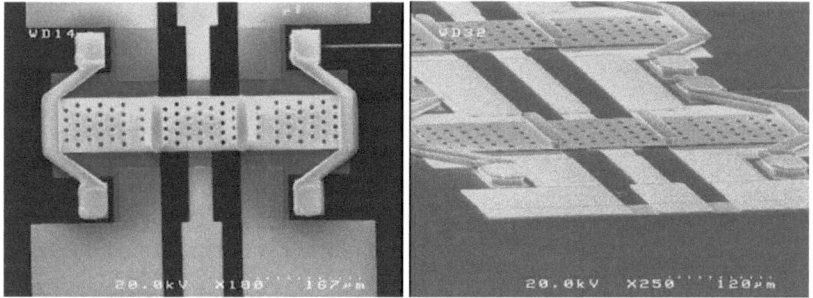

Figure 4.14: SEM images of the tri-state shunt switchable capacitor

### 4.4.2 Electromagnetic Characterisation

Measured S-parameters of the tri-state capacitor are shown in Fig. 4.15. Down-state measurements were done at 45 V, which is 13 % above the pull-in voltage of 40 V. The capacitances extracted with Eq. 3.23 are shown in Fig. 4.16. The transmission line capacitance of 48 fF was subtracted from the resulting values to obtain pure MEMS capacitance. The up-state capacitance is about 25 fF. In the intermediate state, when both the membrane and the CPW center conductor are biased at 45 V, the capacitance is about 114 fF. In the third state, when the both CPW grounds and center conductor are kept at DC-ground, the capacitance is about 400 fF. Thus, a capacitance ratio of 1:4.5:16 is obtained.

### 4.4.3 Electromechanical Characterisation

For this component Laser-Doppler-Vibrometer measurement was not available, thus a theoretical estimation based on the theory described in Section 3.2.3 and confirmed experimentally in Section 4.2 was used. Using Eq. 3.9, the spring constant was extracted to be 150 N/m ($g_0$=2.5 μm, $t_d$=300 nm, $\epsilon_r$=10, $A$=43,000 μm$^2$). From the switch geometry and material density, the mass of the moving part $m$ was calculated to be 1.58 μg. Then the mechanical resonance frequency $\omega_0$ is 308 kHz and the calculated switching time $t_s$ for 45 V ($\delta_2 = 0.09$) is 13.4 μs ( Eq. 3.11 and Eq. 3.14 respectively).

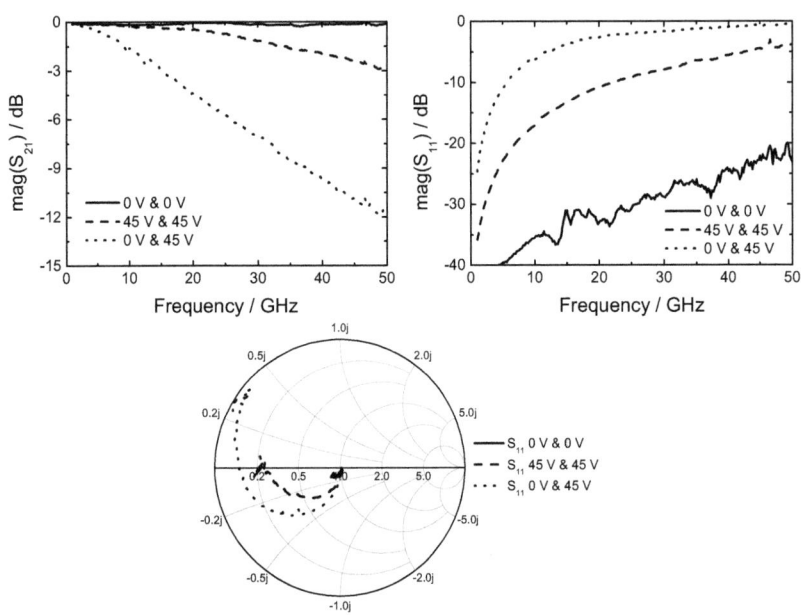

Figure 4.15: Measured S-parameters of the tri-plate capacitor in three bias configurations of the membrane and CPW centre conductor as shown in Fig. 4.13.

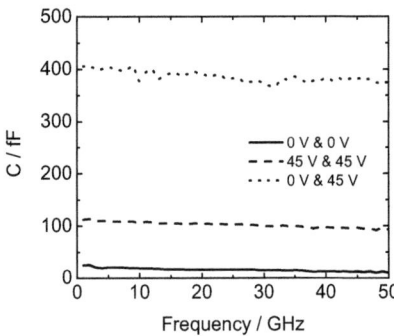

Figure 4.16: Extracted capacitances of the tri-plate capacitor in three bias configurations of the membrane and CPW centre conductor as shown in Fig. 4.13.

## 4.5 Two-state Shunt Switchable Capacitor

### 4.5.1 Design

This and the following switchable capacitors are designed for distributed MEMS transmission line (DMTL) phase shifters and impedance tuners presented in Section 10.2. For periodic-structure applications like DMTL, the requirements for variable capacitors are

- low capacitance ratio, typically between 1.5 and 6

- small lateral dimensions, since they determine the cut-off frequency of the first passband, and thus, useable frequency range of the DMTL circuit

- preferably no DC bias on the RF-signal line to avoid large number of DC-blocks

The switchable capacitor presented in this section has been designed in collaboration with VTT Millilab, Finland. SEM-images of the fabricated structure are shown in Fig. 4.17. The schematic cross-section and equivalent circuit are identical to the ones shown in Fig. 4.6. The same actuation scheme with the bias voltage applied to the membrane was used, since it allows to keep the RF signal line at DC-ground and requires only one DC-control signal per MEMS.

In this switchable capacitor, the low capacitance ratio was achieved by taking advantage of the normally undesired internal stresses, causing lateral bowing of the membrane. Thus, no lateral stiffening bars were used here, in contrast to switches described above, where flat membranes are necessary. The resulting lateral bowing can be observed in the SEM-image shown in Fig. 4.17. As a result, the up-state capacitance is slightly larger than that in case of a flat membrane with stiffening bars. In addition, the down-state capacitance is significantly lower due to reduction of the area, which is in direct contact with the dielectric above the lower electrode. Thus, a low capacitance ratio is achieved, as required by DMTL circuits. It should be noted that this approach is very sensitive to internal stresses in the moveable membrane and is only suitable for well controlled fabrication processes, which may be difficult to achieve in practice.

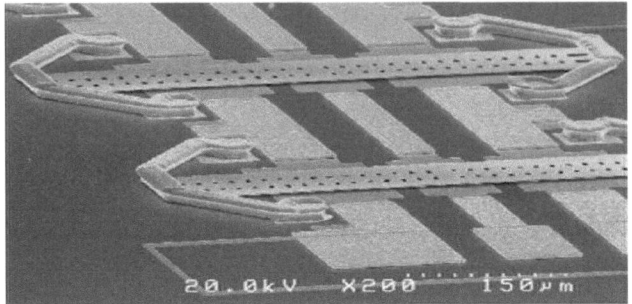

Figure 4.17: SEM images of the two-state shunt switchable capacitor

### 4.5.2 Electromagnetic Characterisation

The switch was characterized up to 50 GHz with the Agilent 8510C vector network analyser using SOLT calibration up to the probe tips. The measurement setup consisted of two GSG probes for the RF-signal and one DC-needle for the bias supply. The dynamic pull-in voltage is about 32 V. The measured performance of the MEMS switch is shown in Fig. 4.18. The extracted capacitance values are about 20 fF in the up-state and 95 fF in the down-state (Eq. 3.23). Thus, the obtained capacitance ratio is 4.75.

### 4.5.3 Electromechanical Characterisation

For this component, the switching time was calculated theoretically using the method from Section 3.2.3, since Laser-Doppler-Vibrometer measurements were not available. The dynamic pull-in voltage of the switch is about 32 V. The down-state measurements were taken at a 9%-higher value of 35 V. The spring constant extracted using Eq. 3.9 is 48 N/m ($g_0$=2.5 µm, $t_d$=300 nm, $\epsilon_r$=10, $A$=20,600 µm²). The mass of the moving part $m$ was calculated from the switch geometry and material density and is equal to 1.04 µg. Finally, the mechanical resonance frequency $\omega_0$ is 216 kHz and the calculated switching time $t_s$ for 35 V ($\delta_2 = 0.09$) is 20.5 µs (Eq. 3.11 and Eq. 3.14).

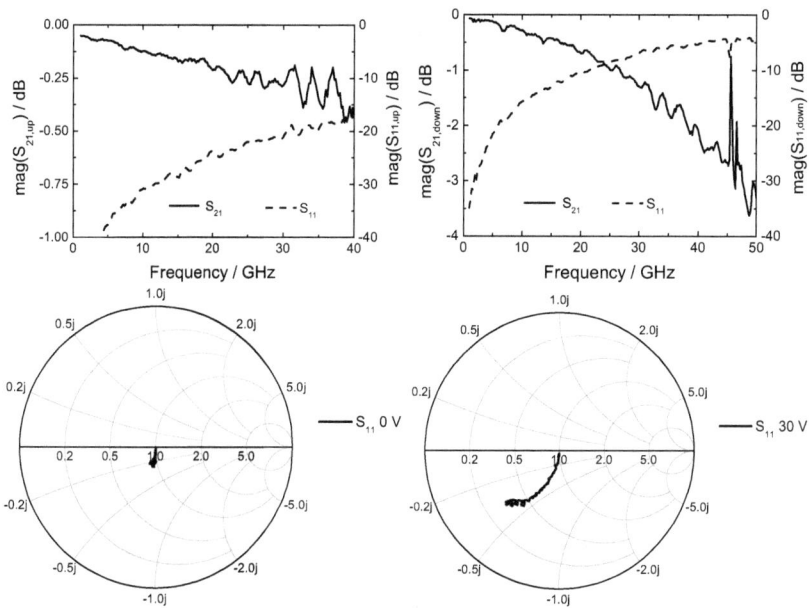

Figure 4.18: S-parameters measured in the up- (left) and down- (right) states of the two-state capacitor shown in Fig. 4.17

## 4.6 Switchable Series Capacitor

### 4.6.1 Design

Switchable series capacitors can be used as building blocks for composite right/left handed (CRLH) transmission lines with "left-handed" properties in certain frequency bands. A CRLH transmission line is a periodic structure, operated at frequencies far away from the passband cut-off, and thus can be treated as a homogenous material, often referred to as "metamaterial". This clearly differentiates metamaterials from e.g. photonic crystals, which, on the contrary, make use of periodic structure stop bands. With MEMS, the "left-handed" frequency band of CRLH lines can be made tunable, enabling new versatile circuits.

The switchable series capacitor presented here is realized as a series cantilever made of 3 µm-thick gold. Originally, no MEMS cantilevers were available at ISiT which could be used

Figure 4.19: SEM image of the series switchable capacitor

as a starting point for the mechanical design. Thus, mechanical simulations were performed using ANSYS software in cooperation with Technion (Israel).

SEM-image and schematic cross-section of the fabricated component are shown in Fig. 4.19 and Fig. 4.20 respectively. The capacitor is realised as an interrupted coplanar line, with RF-coupling due to the cantilever's capacitance between the cantilever's tip and the continuation of the coplanar line, covered with AlN dielectric. The actuation voltage is applied to a separate electrode, which is located below the cantilever and is DC-isolated from the coplanar line. The desired low capacitance ratio was obtained utilizing the internal stress gradient across the electroplated thin-film gold membrane. The stress gradient leads to typically undesired slight upwards bending of the cantilever. However, since there is no electrostatic force acting on the cantilever's tip, the upwards bending preserves also in the down-state. Thus, the cantilever's tip does not touch the CPW center conductor, but there is a small air gap, which results in a low capacitance ratio.

### 4.6.2 Electromagnetic Characterisation

Measured S-parameters for 0 V and 80 V bias are shown in Fig. 4.21. The equivalent circuit is identical to the one shown in Fig. 3.7, indicating that the performance is mainly determined by the series MEMS capacitance. In the up-state the air gap is very large and the MEMS capacitance is very small, about 4 fF (Eq. 3.26). After the actuation voltage is applied, the

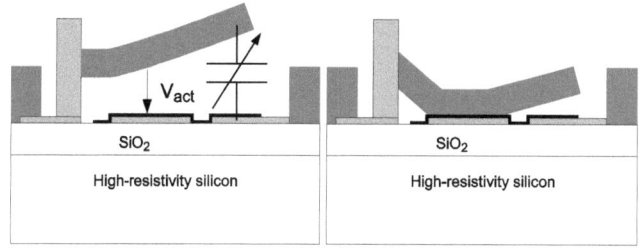

Figure 4.20: Cross-section of the series switchable capacitor shown in Fig. 4.19 in the up- (left) and down-states (right)

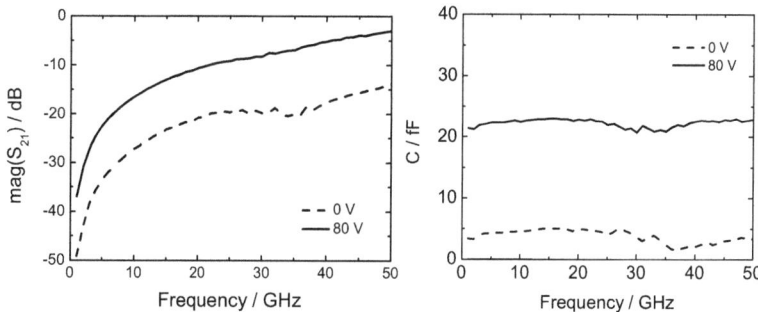

Figure 4.21: Measured transmission coefficient (left) and extracted capacitances (right) in the up- and down states of the series switched capacitor shown in Fig. 4.19

cantilever snaps down on the actuation electrode. The extracted down-state capacitance is 22 fF (Eq. 3.26). The low capacitance ratio of 5.5 is achieved due to the remaining air gap in the down-state. It should be noted that this approach is very sensitive to internal stresses in the cantilever and is only applicable for well controlled fabrication processes, which may be difficult to achieve in practice.

### 4.6.3 Electromechanical Characterisation

The dynamics of the cantilever were measured with a Laser-Doppler-Vibrometer. The laser beam was focused at the tip of the cantilever. To investigate the effect on the actuation voltage on the switching time, 55 V, 80 V, 82 V and 85 V were applied. Fig. 4.22 (left) shows recorded membrane deflections during switching. Measured trajectories agree well with the theoretical ones calculated by D. Elata et. al. in [37–39] (Fig. 3.3). At 55 V, i.e. below the dynamic pull-in, the membrane undergoes oscillations in the region of the air-gap. Above the dynamic pull-in at 80 V, 82 V and 85 V the membrane snaps down to the bottom electrode. Only a 6 % increase in the actuation voltage for 80 V to 85 V reduces the switching time by 84 % from 70 μs to 38 μs.

Note that the maximum deflection of the cantilever tip is 6 μm, and is twice as large as the sacrificial layer thickness (3 μm), defining the air-gap. This is due to the built-in stress gradient, causing the cantilever to bend upwards, as is also seen in the SEM image (Fig. 4.19). Such an increase in the air-gap is the cause of the quite high actuation voltage, as follows from Eq. 3.9.

As the actuation voltage is removed, the release oscillations are observed (Fig. 4.22, right). As for the shunt bridge, the oscillations were fitted with Eq. 3.15 using nonlinear least squares. The extracted resonance frequency is $\omega_d = 110$ KHz and the damping coefficient is $\eta = 15$ KHz. As compared to the shunt bridge design, both values are reduced due to lower spring constant and larger holes in the membrane, which reduce the damping.

To validate the theory described in Section 3.2.3 once again (see Section 4.2), theoretical switching time values are compared to measurements. As explained above, the dynamic pull-in voltage of the switch is about 80 V and measurements were taken for 82 V and 85 V. Using Eq. 3.9, the spring constant was extracted to be 9 N/m ($g_0$=6 μm, $t_d$=300 nm, $\epsilon_r$=10, $A$=7,500 μm²). From the switch geometry and material density, the mass of the moving part $m$ was calculated to be 0.64 μg. Then using Eq. 3.11 and Eq. 3.14, the mechanical resonance frequency $\omega_0$ is 108 kHz, which is quite close to 110.2 kHz, calculated using Eq 3.15 and experimental values of $\omega_d$=110 kHz and $\eta$=15 kHz, extracted from the release measurement shown in Fig. 4.22. The calculated switching time $t_s$ for 82 V ($\delta_2 = 0.02$) is 49.2 μs and for 85 V ($\delta_2 = 0.06$) is 42.8 μs.

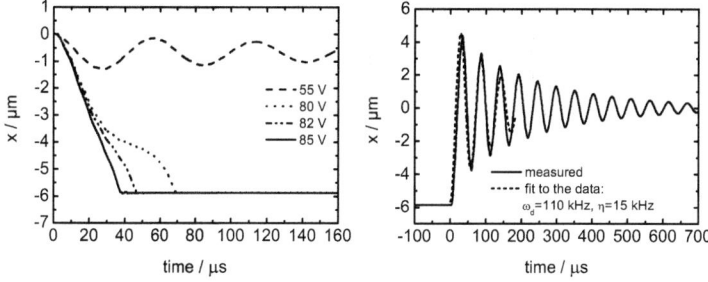

Figure 4.22: Switching and release of the series switchable capacitor shown in Fig. 4.19

## 4.7 Cantilever-based Shunt Switchable Capacitor

### 4.7.1 Design

The main motivation behind this design was to develop a component, suitable for DMTL circuits for upper millimeter-wave frequencies. As described above, bridges realised using thin Au-Ni-Au membrane have Ni supporting springs for reliable operation. These springs increase the MEMS width by up to 50%, which reduces the passband cut-off frequency of periodic structures, as analytically shown in Part II. Thus, for high frequency operation of circuits, lateral dimensions of the switched capacitor must be reduced.

SEM-images of the fabricated structure are shown in Fig. 4.23. This switchable capacitor was designed by placing two cantilevers from Section 4.6 in shunt to mimic a bridge design. Thus, the width of such a capacitor is only 60 µm, in contrast to 330 µm needed for a true "bridge"-type design described in Section 4.4. Besides, since these cantilevers are made up of thick gold, a lower loss can be achieved as compared to the thin Au/Ni membrane.

The cross-section is schematically illustrated in Fig. 4.24. As in the series cantilever design, the low capacitance ratio is due to MEMS upwards bending and thus, a remaining air-gap in the down-state. The bias voltage is applied directly to the MEMS cantilevers, and the underlying coplanar grounds and center conductor are used as actuation electrodes.

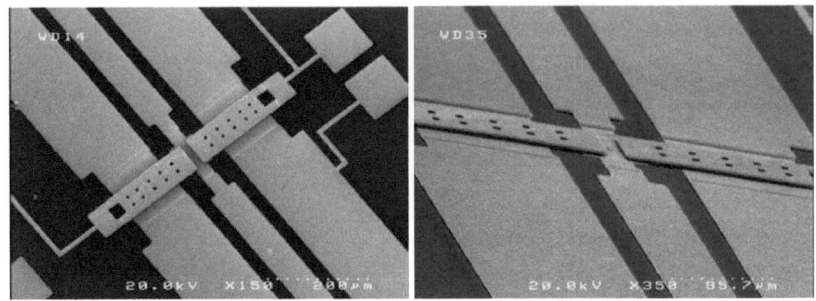

Figure 4.23: SEM images of the cantilever-based shunt switchable capacitor

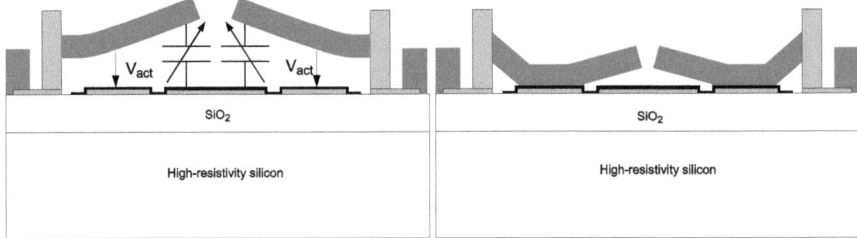

Figure 4.24: Cross-section of the cantilever-based shunt switchable capacitor shown in Fig. 4.23. Left: up-state, right: down-state.

### 4.7.2 Electromagnetic Characterisation

Measured S-parameters are shown in Fig. 4.25 and extracted capacitance is shown in Fig. 4.26. The equivalent circuit of this design is identical to that of the shunt MEMS bridge shown in Fig 4.6 (right). The overall shunt capacitance is equal to 14 fF in the up-state and 150 fF in the down-state (Eq. 3.23). The capacitance ratio to about 10.7 and is about twice as high than for the series switchable capacitor. This increase is due to the attractive force acting on the tip of the cantilever, since the coplanar line is kept at DC-ground, resulting in a reduced remaining air-gap in the down-state, increasing the down-state capacitance.

### 4.7.3 Electromechanical Characterisation

The electromechanical performance of this component is identical to that of the series cantilever properties, described in Section 4.6.

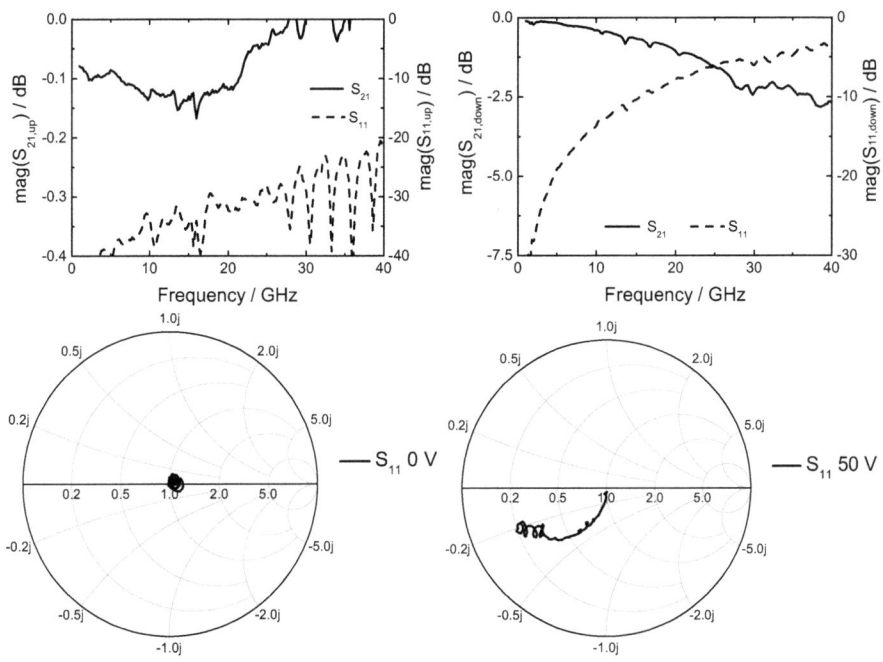

Figure 4.25: Measured S-parameters of the cantilever-based shunt capacitor shown in Fig. 4.23 in the up- (left) and down-states (right)

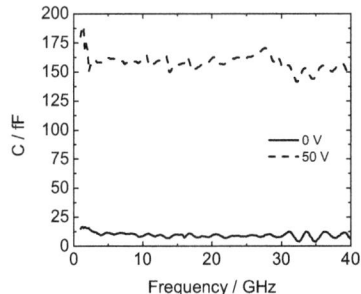

Figure 4.26: Extracted capacitance of the cantilever-based shunt capacitor

# Chapter 5

# Components in FBK RF-MEMS Technology

This chapter presents components developed using the RFMEMS technology of the Fondazione Bruno Kessler, located in Trento, Italy. This technology undergoes a constant development and is not yet commercially available. The components were developed within the AMICOM project cooperation.

## 5.1 FBK RFMEMS Fabrication Process

As in case of the Fraunhofer ISiT, the RFMEMS fabrication process at FBK is based on a surface-micromachining. Similar to the ISiT RFMEMS technology, oxidised high-resistivity silicon is used as the substrate material. Main distinguishing features of the FBK-process are gold membranes either 1.8 µm or 4.8 µm thick and an additional Cr/Au metal layer, which can be used for metal-to-metal contact switches and for capacitive switches with a "floating" metal contact. Use of the floating metal potentially results in a well defined down-state capacitance, which is independent on the dielectric surface roughness. Since the silicon surface is not treated e.g. with poly-Si or Argon prior to oxidation (as is the case in ISiT), formation of a conducting channel at the $Si/SiO_2$ interface is expected.

Main RFMEMS fabrication steps are illustrated in Fig. 5.1 and described in details below:

1. Growth of a 1 µm thick silicon oxide on top of high-resistivity silicon substrate

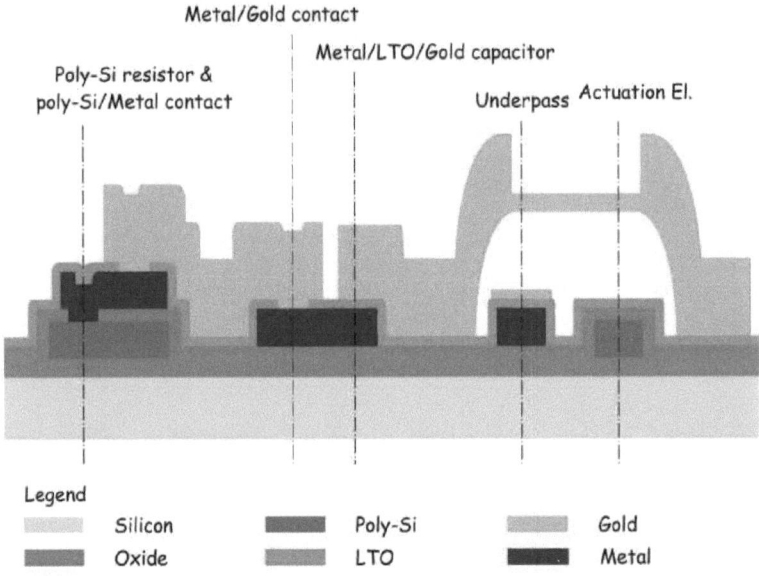

Figure 5.1: Schematic illustration of the FBK RFMEMS technology (taken from [51])

2. Deposition of 630 nm thick poly-silicon for high-resistivity bias lines and actuation electrodes

3. Growth of 300 nm thermal oxide for DC-isolation of bias electrodes

4. Deposition and definition of Ti/TiN/Al/Ti/TiN 30/50/410/30/80 nm multilayer metal (underpass)

5. Deposition of 100 nm thick LPCVD oxide (LTO) to form the MEMS capacitance

6. Floating metal deposition (10/150 nm Cr/Au ) used as "floating" metal for controlled down-state capacitance and for metal-to-metal contact switches

7. Deposition and definition of 3 μm thick resist spacer to define the air-gap

8. Electroplating of the first Au layer (1.8 μm thick bridges and pads)

9. Electroplating of the second Au layer (3 μm thick CPW lines)

10. Final MEMS release with wet etching and critical-point drying

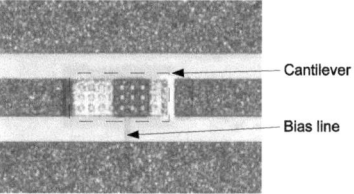

Figure 5.2: Micrograph of the series metal-to-metal contact cantilever switch

## 5.2 Series Metal-to-Metal Contact Switch

### 5.2.1 Design

The series metal-to-metal contact switch was designed as a coplanar line with a gap in the center conductor, which can be closed with a MEMS cantilever, as can be seen in the micrograph in Fig. 5.2. The cantilever design is based on that reported by P. Farinelli in [52], but has been optimized to be used in RFMEMS switchable defected-ground structures presented in Section 5.3.

An SEM image of the cantilever is shown in Fig. 5.8, where it is used as part of an RFMEMS-switchable inductance, described in section Section 5.3. The cantilever's cross-section is schematically illustrated in Fig. 5.3. The moveable membrane is fabricated from the thinner gold layer (1.8 µm), and is reinforced in the central part with the second gold layer to a total thickness of of 4.8 µm. Such local membrane reinforcement together with dimples are necessary for a high contact force and thus, low contact resistance in the down-state [52]. The high-voltage actuation electrode is covered with 300 nm of $SiO_2$ to prevent any DC-short circuit in the down-state.

### 5.2.2 Electromagnetic Characterisation

The dynamic pull-in voltage of the switch was experimentally determined to be 56 V. Measured S-parameters for the up- and down-states are shown in Fig. 5.4. In the up-state, the cantilever equivalent circuit corresponds to the one shown in Fig. 3.7 (left), and the performance is mainly determined by the series up-state capacitance. For this design, the

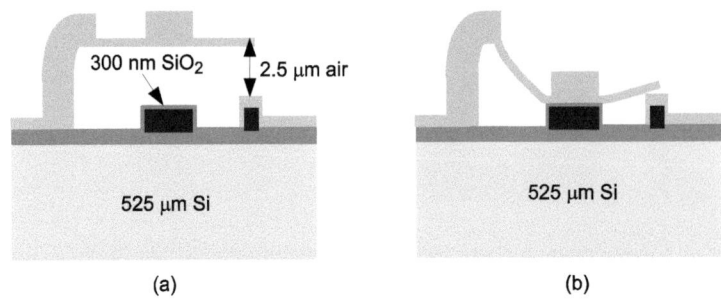

Figure 5.3: Schematic cross-section of the series switch shown in Fig. 5.2 in the up- (left) and down- (right) states (not to scale)

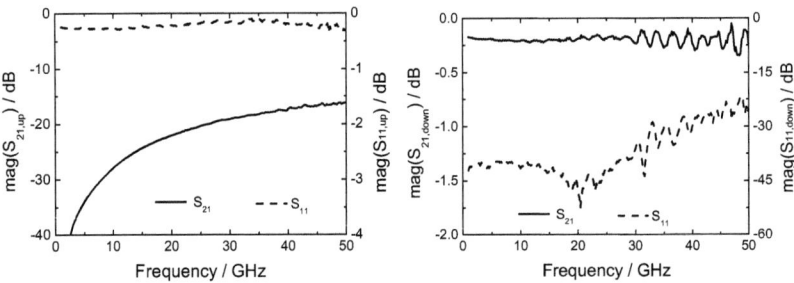

Figure 5.4: Measured S-parameters of the series cantilever switch shown in Fig. 5.2 in the up- (left) and down- (right) states

up-state capacitance was extracted to be 6.5 fF (Eq. 3.26). The corresponding isolation is better than 15 dB up to 50 GHz.

In the down-state, the equivalent circuits is given in Fig. 3.7 (right). The performance is now mainly determined by the contact resistance, which is about 2 $\Omega$ (Eq. 3.27). As a result, the insertion loss of the switch is better than 0.3 dB up to 50 GHz.

### 5.2.3 Electromechanical Characterisation

Fig. 5.5 shows switching and release dynamics of the cantilever, measured with a Laser-Doppler-Vibrometer. The switching time reduces from 58 µs to 28 µs as the actuation voltage is increased from 56 V to 70 V. As before, the release oscillations were fitted with the

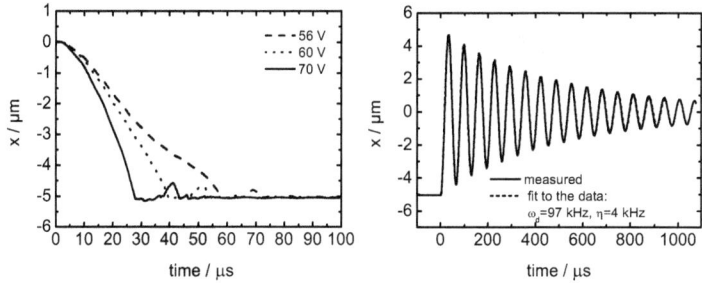

Figure 5.5: Dynamic deflection of the series ohmic switch shown in Fig. 5.2 during switching and release measured with Laser Doppler Vibrometer

non-linear squares algorithm. The extracted mechanical parameters are $\omega_d = 97$ kHz and $\eta = 4$ kHz. The mechanical resonance frequency of the FBK cantilever is close to that of the ISiT cantilever, since they both are made of gold of similar thickness and also have similar geometries, thus, similar spring constants and masses.

As for the ISiT cantilever, the theory described in Section 3.2.3 has been validated by comparing theoretical switching time values with the measured ones. As explained above, the dynamic pull-in voltage of the switch is about 56 V and measurements were taken for 60 V and 70 V. Using Eq. 3.9, the spring constant was extracted to be 6.5 N/m ($g_0$=5 µm, $t_d$=300 nm, $\epsilon_r$=4, $A$=7,700 µm²). From the switch geometry and material density, the mass of the moving part $m$ was calculated to be 0.68 µg. Then using Eq. 3.11 and Eq. 3.14, the mechanical resonance frequency $\omega_0$ is 98 kHz, which is very close to 97 kHz calculated using Eq 3.15 and experimental values of the damped mechanical oscillation frequency of 97 kHz and the damping coefficient $\eta$ = 4 kHz extracted from the release measurement shown in Fig. 5.5. The calculated switching time $t_s$ for 60 V ($\delta_2 = 0.07$) is 48 µs and for 70 V ($\delta_2 = 0.25$) is 35.2 µs, which is in good agreement with the measured values.

## 5.3 RFMEMS-Switchable Series Inductance using Defected Ground Structures

### 5.3.1 Design

Most of RFMEMS reported in the literature produce a changing capacitance. However, some applications like DMTL phase shifters, tunable filters, tunable matching networks and other may require tunable inductors, which are not readily available. Here, switchable defected ground structures (DGS) were used to realize RFMEMS-switchable inductors for micro- and millimeter wave frequencies.

A defected ground structure is obtained by introducing a cut-out in the ground plane of a transmission line. It is modelled as a parallel LCR resonance circuit in series to the transmission line [53]. When operated well below its LC resonance frequency, a DGS acts as a series inductance and is often used to introduce a slow-wave effect in transmission lines [54]. At the resonance frequency, the DGS represents a very high series impedance, causing total signal reflection and is sometimes used in filters. Above the resonance frequency, a DGS is equivalently a series capacitance.

In this work, a coplanar defected ground structure was made switchable by means of two metal-to-metal contact RFMEMS cantilevers described in Section 5.2. A micrograph and an SEM-image of the fabricated MEMS-DGS structure are shown in Fig. 5.6 (left) and Fig. 5.8 respectively. As the photos illustrate, the DGS is realized by two rectangular cut-outs in the coplanar ground plane, whose shape was optimised with EM-simulation (Sonnet). The cut-outs can be shorted or open upon actuation of two MEMS cantilevers.

The equivalent circuit of the MEMS-DGS structures is shown in Fig. 5.6 (right). When the cantilever MEMS switches are open, the structure behaves as a series inductance since the DGS is operated well below its resonance frequency. When the MEMS cantilevers are actuated, they short-circuit the inductance, resulting in a coplanar line with continuous ground structures. To verify the functionality of this approach, a current distribution was calculated with EM-simulation (Sonnet) assuming an ideal ohmic contact in the down-state. The result is shown in Fig. 5.7, indicating a longer path for the current when the cantilever is in the up-state, thus increased series inductance. In the cantilever's down state, the current

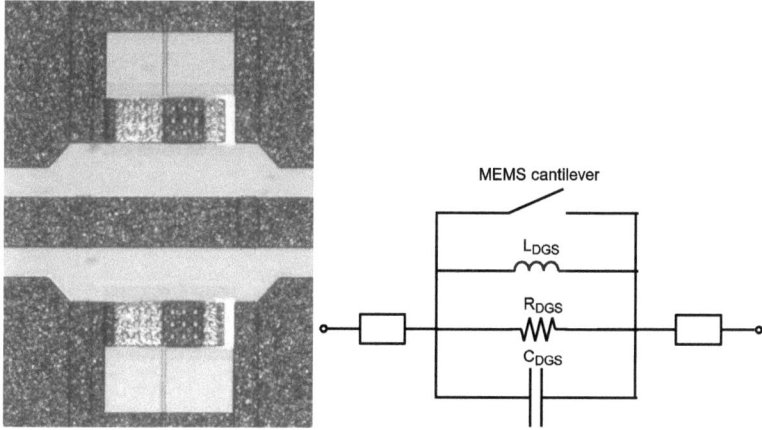

Figure 5.6: Micrograph (left) and equivalent circuit (right) of the defected ground structure switchable with RFMEMS switches from Fig. 5.2

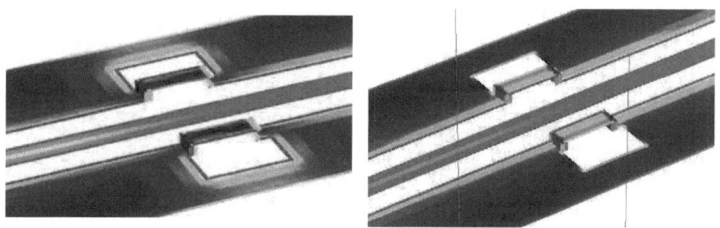

Figure 5.7: DGS current distribution simulated with Sonnet in the up- (left) and down- (right) states of the MEMS switches

does not "see" the defected ground structure and flows on the very edges of the coplanar ground planes, as if the line were uniform.

### 5.3.2 Electromagnetic Characterisation

Measured S-parameters for the up- and down-states of MEMS cantilevers are shown in Fig. 5.9. The behaviour as a switchable inductance is best observed by studying the reflection coefficient in the Smith Chart and by extracting $L$ from the imaginary part of the inverse $Y_{21}$ parameter:

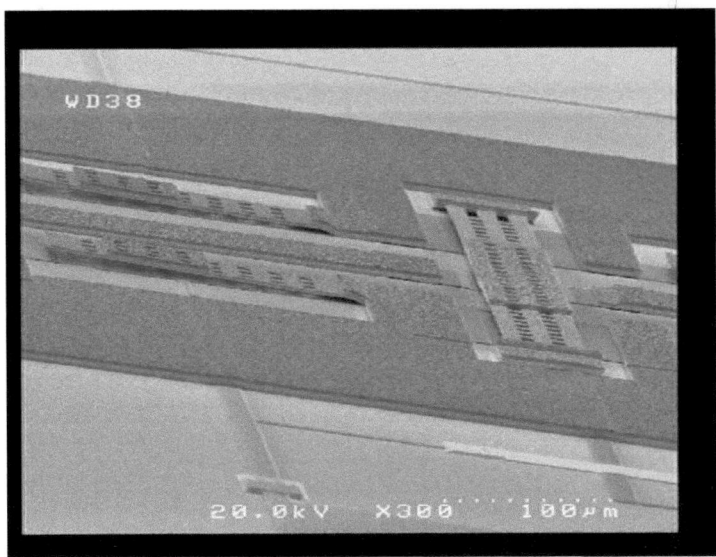

Figure 5.8: SEM image of the defected ground structure switchable with MEMS switches from Fig. 5.2 and the shunt capacitor presented in Section 5.4. This structure is also used as a unit cell of the DMTL phase shifter described in Section 9.1.

$$L = -\frac{imag(\frac{1}{Y_{21}})}{\omega}$$

The calculated inductances are plotted in Fig. 5.10. In the down-state, the structure behaves like a series inductance of about 135 pH. This is solely due to the impedance of the unloaded transmission line, which was chosen slightly larger than 50 $\Omega$ to compensate for up-state capacitances of MEMS bridges in the subsequent DMTL integration, as explained in Section 10.1 and Section 9.1. In the up-state, the inductance is increased to about 205 pH due to an additional DGS inductance. The resulting $L$ of the MEMS defected ground structure is about 70 pH.

### 5.3.3 Electromechanical Characterisation

Electromechanical characterisation of the DGS switchable inductance is identical to that of the cantilever, described in Section 5.2.3

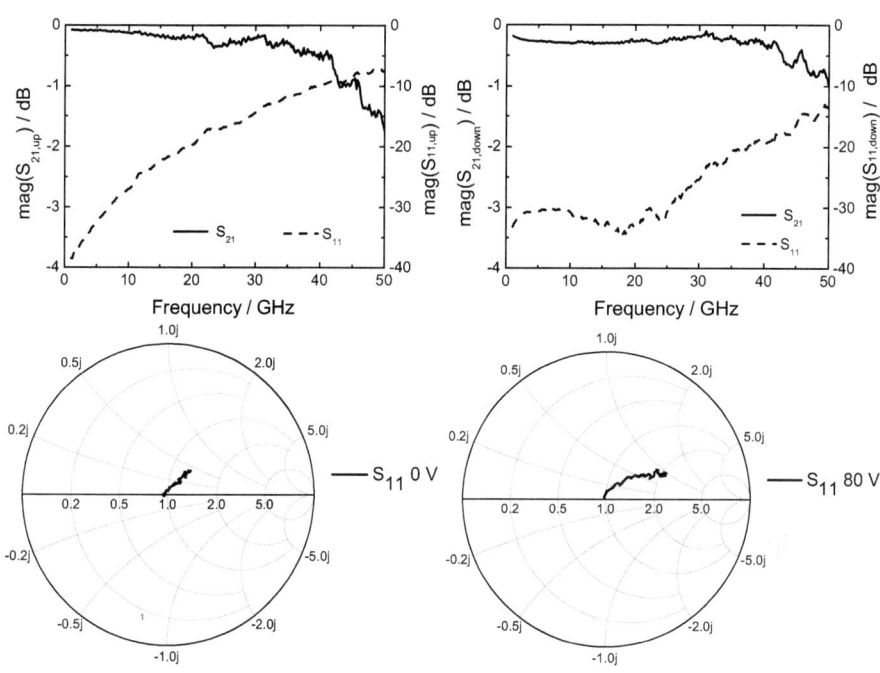

Figure 5.9: Measured S-parameters of the RFMEMS-switchable defected ground structure in the up- (left) and down- (right) states

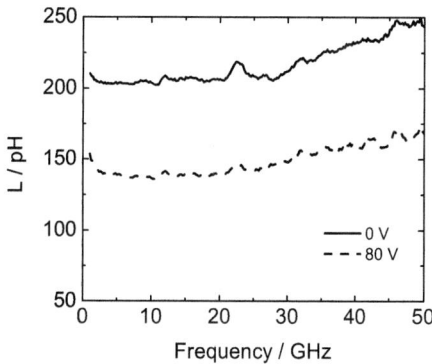

Figure 5.10: Extracted inductances of the RFMEMS-switchable inductance in the up- and down-state

## 5.4 Shunt Switchable Capacitor

### 5.4.1 Design

The shunt switchable capacitor is made of a MEMS bridge suspended over a coplanar line covered with a dielectric layer. The bridge and the underlying coplanar line act as a shunt capacitor to ground and in the down-staten the capacitance value is relatively high. However, a low down-state capacitance is required for switchable capacitor operation. Thus, an intentional air-gap in the down-state was introduced to lower the capacitance ratio, similar to the component in ISiT technology from Section 4.2.

A microphotograph and an SEM-image of the fabricated structure are shown in Fig. 5.11 and Fig. 5.8 respectively. The schematic cross-section is illustrated in Fig. 5.12. The suspended membrane is made of a thin gold layer (1.8 µm) and has local membrane reinforcement with 3 µm of gold. The intentional air-gap in the down-state was created by thinning the oxide layer to recess the centre conductor into it. Due to the local membrane reinforcement, the membrane's movement is stopped by the ground planes and in the down-state the central part remains flat and does not bend towards the recessed centre conductor. Thus, in the down-state a 500 nm air gap remains, ensuring a theoretical capacitance ratio of 6.

The disadvantage of this component is that it uses a very thin dielectric layer (100 nm) for DC-isolation of the actuation area. It was revealed in the experimental characterisation that this switchable capacitor can be suspected to dielectric charging, leading to sticking. This problem can be solved by redesigning the component in a way that the thicker (300 nm) dielectric layer is used for DC-isolation, as is the case in the cantilever switch presented in Section 5.2.

### 5.4.2 Electromagnetic Characterisation

Measured S-parameters taken at 0 and 34 V, which is slightly above the dynamic pull-in, are shown in Fig. 5.13. The equivalent circuit of this component is the same as shown in Fig. 4.6 (right). The dominating parameter is the MEMS capacitance, which was extracted to be 31 fF in the up-state and 125 fF in the down-state (Eq. 3.23). Thus, the measured capacitance ratio is 4 and not 6 as predicted theoretically. It can be explained by some

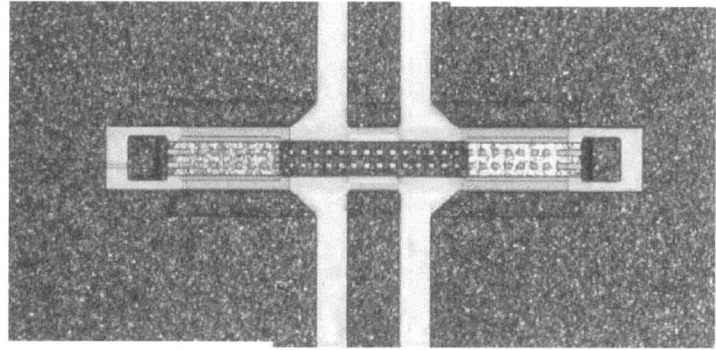

Figure 5.11: Micrograph of the shunt switchable capacitor presented in in Section 5.4. DC-bias pad is not shown.

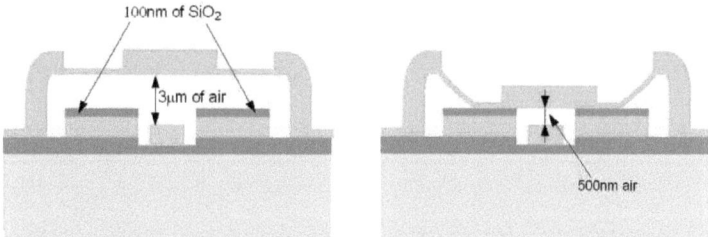

Figure 5.12: Schematic cross-section of the shunt switchable capacitor presented in Section 5.4 in the up- (left) and down- (right) states

downwards bending of the membrane in the up-state, increasing the up-state capacitance, while having little effect in the down-state.

### 5.4.3 Electromechanical Characterisation

Membrane deflections during switching and release are shown in Fig. 5.14. As before, the trajectories were obtained with a Laser-Doppler-Vibrometer, focusing the beam on the bridge central point. The measurement was performed for 50 V. The corresponding switching time is 35 μs, and it can be decreased at the expense of increased actuation voltage. Fitting of the release oscillations resulted in the mechanical resonance frequency $\omega_d = 149$ kHz and the damping coefficient $\eta = 8$ kHz.

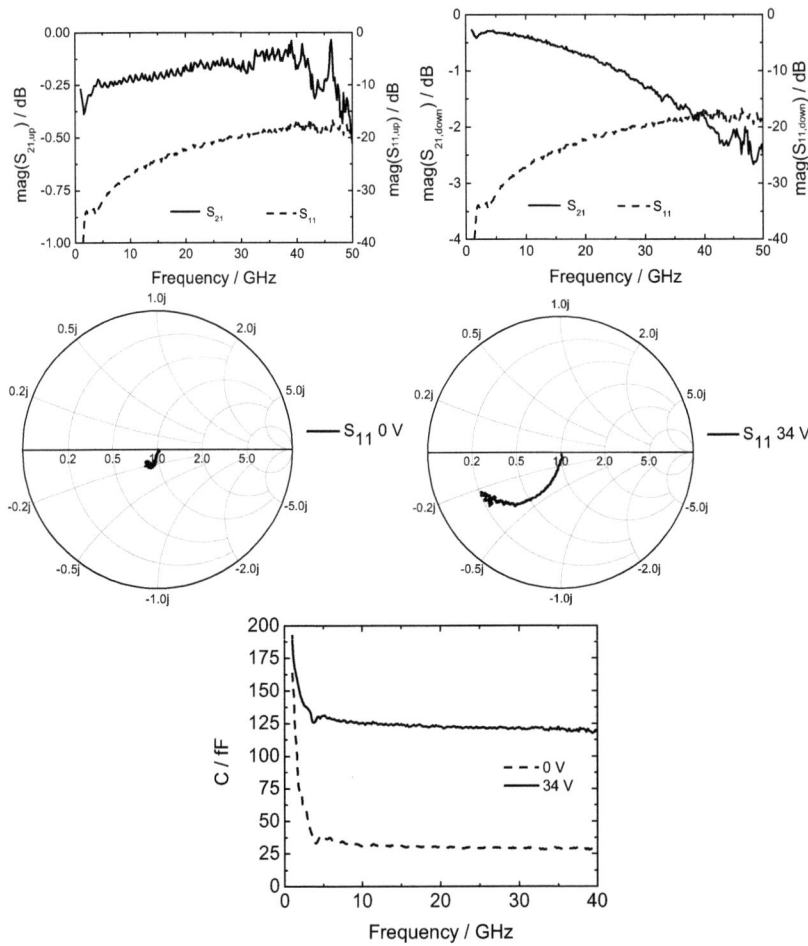

Figure 5.13: Measured S-parameters and extracted capacitances of the switchable shunt capacitor in the up- (left) and down- (right) states

Once again, the theory described in Section 3.2.3 has been validated here by comparing theoretical switching time values with the measured ones. As explained above, the dynamic pull-in voltage of the switch is about 45 V and the LDV measurement was taken for 50 V. Using Eq. 3.9, the spring constant was extracted to be 35 N/m ($g_0$=3 µm, $t_d$=100 nm, $\epsilon_r$=4, $A$=13,700 µm²). From the switch geometry and material density, the mass of the moving

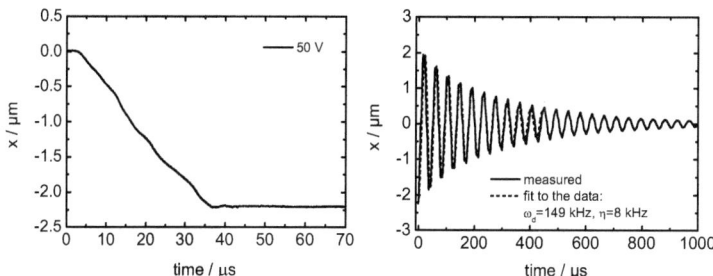

Figure 5.14: Dynamic deflection of the shunt switchable capacitor during switching and release measured with LDV

part $m$ was calculated to be 1.56 μg. Then using Eq. 3.11 and Eq. 3.14, the mechanical resonance frequency $\omega_0$ is 150 kHz. This is quite close to 149.05 kHz, calculated using Eq 3.15 and experimental values of the damped mechanical oscillation frequency of 149 kHz and the damping coefficient $\eta$=8 kHz, extracted from the release measurement shown in Fig. 5.5. The calculated switching time $t_s$ for 50 V ($\delta_2 = 0.11$) is 28 μs and this value is quite close to the measured one.

# Part II

# Modelling of RFMEMS Periodic Structures

# Chapter 6

# Introduction

The most commonly used RFMEMS-tunable periodic structures are distributed MEMS transmission lines (DMTL). DMTL impedance tuners and phase shifters with excellent performance have been reported (e.g. [1, 3]). The applications for these circuits are diverse: tunable matching networks for reconfigurable active circuits, impedance tuners for noise- and load-pull measurements, phase shifters for beam-steerable phased arrays and other. Thus, these RFMEMS periodic structures can be classified as one-dimensional structures for guided waves normally operated in the frequency passbands. Their properties, modelling and applications is the main focus of this thesis.

The most common realization of such circuits is a transmission line periodically loaded with shunt RFMEMS capacitors. Its unit cell is formed by a tunable or switchable MEMS capacitor $C$ and two sections of transmission line of length $d/2$ each and characteristic impedance $Z_0 = \sqrt{L'/C'}$ ($L'$, $C'$ are per unit length inductance and capacitance of the unloaded transmission line), as schematically shown in Fig 6.1.

A common modelling approach is to restrict its scope to a low-frequency region, where the structures can be treated as smooth artificial transmission lines with an equivalent characteristic impedance $Z_{DMTL}$ and an equivalent phase velocity $v_{ph,DMTL}$ given by the following equations:

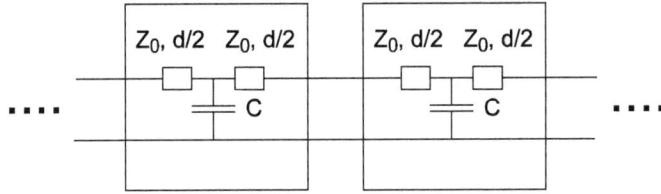

Figure 6.1: Transmission line with characteristic impedance $Z_0$ loaded with shunt capacitors $C$ after each period $d$

$$\begin{cases} Z_{DMTL} = \sqrt{\dfrac{L'}{C' + C/d}} \\ v_{ph,DMTL} = \sqrt{\dfrac{1}{L'(C' + C/d)}} \end{cases} \quad (6.1)$$

This approximation is accurate at low frequencies, i.e. if $d \ll \lambda$. However, it neglects the frequency dispersion arising due to the periodic structure and leads to a large deviation from the real case as frequency increases. Unfortunately, often DMTL circuits are operated at frequencies where the low frequency approximation is not valid and characteristic impedance and propagation constant significantly deviate from their low-frequency values due to dispersion. Thus, a full broadband analysis is unavoidable and is described in the following sections.

In addition, accurate estimation of the pass- and stopband cut-off frequencies is essential for circuit design. In the conventional simplified DMTL model, the cut-off frequency of only the first passband is derived. For that, equivalent model consisting of a lumped inductor $L'd$ and a lumped capacitor $C'd + C$ is studied. It has been shown that the cut-off frequency of such a structure is given by Eq. 6.2 [55]. Due to the similarity of the physical background of stopbands to Bragg reflections, this frequency is often referred to as Bragg frequency in the literature [55].

$$f_{Bragg} = \dfrac{1}{\pi d \sqrt{L'(C' + \frac{C}{d})}} \quad (6.2)$$

However, since equation 6.2 is based on a lumped approximation, it tends to underestimate the cut-off frequency and thus, more accurate analysis is needed. An important contribution to the modelling of RFMEMS periodic structures has been made in [4]. Using on the methodology proposed in [42], reference [4] derives closed-form equations for impedance and propagation constant of CRLH periodic structures, i. e. with series capacitive and shunt inductive loading.

This work deals with right-handed structures and covers in detail the commonly-used shunt capacitive loading as well as series inductive loading and combination of both shunt capacitive and series inductive loading. Simple, accurate and frequency-dependent equations for the propagation constant and equivalent characteristic impedance are obtained by adopting the "loading factor" concept used in [5]. Besides, accurate equations for cut-off frequencies of all pass- and stopbands are derived for shunt capacitive or series inductive loading cases.

# Chapter 7

# Theoretical Background

## 7.1 Space Harmonics Analysis of General 1-D Periodic Structures

Dispersion characteristics of a periodic structure can be tailored by a proper design of its unit cell. To discuss how to obtain a periodic structure with desired properties at a particular frequency range, the underlying theory is briefly reviewed.

It can be shown ( [7, 42, 56, 57]) that the electric field in a periodic structure with a period $d$ can be generally written as:

$$E(z) = e^{\gamma z} P(z) \qquad (7.1)$$

where $P(z)$ is a periodic function with the same period $d$ and $\gamma = -\alpha + j\beta$ is an effective propagation constant with $\beta$ and $\alpha$ being phase and attenuation constants respectively. In the literature, Eq. 7.1 is called Floquet's theorem, which indicates that at any plane $z = md$ ($m$ an integer) the field distribution in a periodic structure repeats itself, except for the propagation factor $e^{\gamma d}$. The function $P(z)$ in Eq. 7.1 can be rewritten using a Fourier series to obtain:

$$E(z) = e^{\gamma z} \sum_{n=-\infty}^{\infty} c_n e^{j\frac{2\pi n}{d} z} \qquad (7.2)$$

or equivalently:

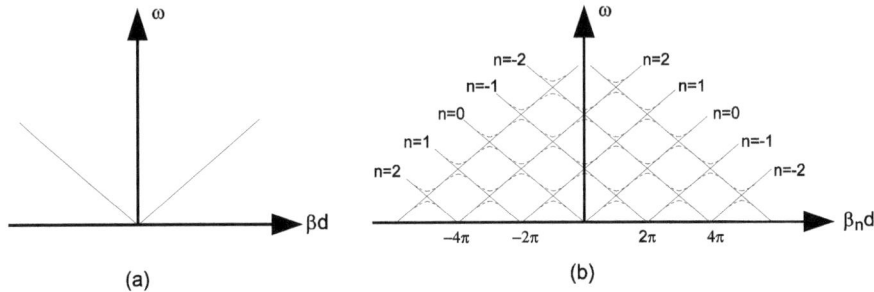

Figure 7.1: Schematic Brillouin-diagram for wave propagation in a homogeneous medium (a) and in a periodic structure, resulting in an infinite number of spatial harmonics (b)

$$E(z) = e^{-\alpha z} \sum_{n=-\infty}^{\infty} c_n e^{j\frac{(\beta d + 2\pi n)}{d} z} \qquad (7.3)$$

Eq. 7.3 represents the electric field as an infinite sum of space harmonics. The $n^{th}$ harmonic is characterized by its phase constant $\beta_n = \frac{\beta d + 2\pi n}{d}$ and a generally complex amplitude coefficient $c_n$. Individual space harmonics do not exist in the structure on their own and are only a mathematical way to represent the wave in a periodic structure. However, such decomposition of the field helps to reveal dispersion characteristics of periodic structures, which are analyzed qualitatively in the following.

Dispersion properties are often studied by means of a Brillouin-diagram illustrating the dependence of the frequency $\omega$ and phase constants of space harmonics $\beta_n$. First, a homogeneous medium with an effective phase constant $\beta$ is considered. The corresponding $\omega - \beta$ diagram is given by two straight lines, as shown in Fig. 7.1 (a). Using this result, individual phase constants of space harmonics $\beta_n$ in a periodic structure are obtained by shifting the $\beta$ curve by $\frac{2\pi n}{d}$, as follows from Eq. 7.3. The resulting Brillouin-diagram combining all space harmonics is schematically shown in Fig. 7.1 (b).

From the diagram it can be seen that each space harmonic has its own phase velocity $v_{ph,n} = \omega/\beta_n$. However, space harmonics with larger $\beta_n$ have proportionally smaller phase velocity and vice versa. Consequently, all space harmonics have the same group velocity $v_{g,n} = \partial \omega / \partial \beta_n$, as is also clear from the diagram [56]. Thus, the effective phase constant of the whole wave $\beta$ is the same as the phase constant of the fundamental frequency. Besides, half of the space harmonics are forward waves (phase and group velocities are co-directional) and half are backward waves (phase and group velocities are counter-directional).

Fig. 7.1 (b) illustrates that whenever $\beta_n d = m\pi$, phase velocities of $n^{th}$ and $-(n+m)^{th}$ harmonics are equal, but their group velocities have opposite signs. Besides, it will be shown in Eq. 7.7 that the amplitudes of these harmonics are also equal at $\beta_n d = m\pi$. This leads to pair-wise destructive interference and thus to a standing wave, prohibiting propagation. This effect is similar to Bragg reflections in optics. At these frequencies the group velocity goes to zero, leading to multiple stopbands around these points. Hence, the resulting dispersion characteristics vary strongly from that in the homogenous medium and are schematically indicated by dashed lines in Fig. 7.1 (b). Exact shapes of $\beta_n$ for different RFMEMS periodic structures will be calculated in Chapter 8.

It is shown in Section 7.2.3 that within stopbands the effective phase constant $\beta d$ equals $m\pi$ and that the effective propagation constant $\gamma$ also has a nonzero real part $-\alpha$. Outside of stopbands, i.e. in the passbands, the effective propagation constant $\gamma$ is purely imaginary ($\alpha = 0$) and is given by the phase constant of the fundamental harmonic. This is summarized in the following equation [57]:

$$\gamma d = \begin{cases} jm\pi - \alpha d, & \text{for } m^{th} \text{ stopband} \\ j\beta d, \ \beta d \in \{m\pi, (m+1)\pi\}, & \text{for } (m-1)^{th} \text{ passband} \end{cases} \quad (7.4)$$

Using Eq. 7.4 for stopband regions, Eq. 7.3 can be rewritten as [57]:

$$E(z) = e^{-\alpha z} e^{\frac{jm\pi}{d}z} \sum_{n=-\infty}^{\infty} c_n e^{\frac{j2\pi n}{d}z} = e^{-\alpha z} \hat{P}(z) \quad (7.5)$$

where $\hat{P}(z)$ is:

$$\hat{P}(z) = e^{\frac{jm\pi}{d}z} \sum_{n=-\infty}^{\infty} c_n e^{\frac{j2\pi n}{d}z} =$$
$$= c_0 e^{\frac{jm\pi}{d}z} + c_1 e^{\frac{jm\pi+2\pi}{d}z} + c_2 e^{\frac{jm\pi+4\pi}{d}z} + c_{-1} e^{\frac{jm\pi-2\pi}{d}z} + c_{-2} e^{\frac{jm\pi-4\pi}{d}z} + \ldots \quad (7.6)$$

$\hat{P}(z)$ is a periodic function in $z$ with period $d$ or $2d$ for even or odd values of $m$, respectively. It can be shown [58] that $\hat{P}(z)$ is either purely real or purely imaginary, but not complex.

Hence, using Euler's formula one notices that the coefficients $c_n$ must fulfill the following conditions:

$$\begin{cases} |c_n| = |c_{-(n+m)}| \\ \arg(c_n) = \begin{cases} \arg(c_{-(n+m)}), & \text{if } \hat{P}(z) \text{ is real} \\ \arg(c_{-(n+m)}) + \pi, & \text{if } \hat{P}(z) \text{ is imaginary} \end{cases} \end{cases} \quad (7.7)$$

In the infinite sum given by Eq. 7.6 the coefficients $c_n$ and $c_{-(n+m)}$ are in front of an exponent and its corresponding inverse. Thus, the conditions of Eq. 7.7 indicate a pair-wise destructive interference of all space harmonics. This results in a stopband, as was also discussed above in the analysis of the Brillouin diagram in Fig. 7.1.

## 7.2 ABCD-Analysis of General 1-D Periodic Structures

Although the analysis of space harmonics based on Floquet's theorem reveals important physical insights, the associated equations can be very involved. On the contrary, the ABCD-parameters approach leads to in a much simpler mathematical description of periodic structures and allows to readily derive dispersion characteristics and other important parameters.

### 7.2.1 Periodic Structures with Finite Number of Sections

To find the dispersion characteristic of a periodic structure, we derive the equation for propagation constants using ABCD-analysis methods. The following discussion partly follows [42] and [59]. A general periodic structure can be represented by a cascade of identical unit cells, as schematically shown in Fig. 7.2. The ABCD parameters of a unit cell are defined by the following equation, with current and voltage arrangement as indicated in Fig. 7.2:

$$\begin{bmatrix} V_n \\ I_n \end{bmatrix} = \begin{bmatrix} A & B \\ C & D \end{bmatrix} \begin{bmatrix} V_{n+1} \\ I_{n+1} \end{bmatrix} \quad (7.8)$$

According to Floquet's theorem (Eq. 7.1), if the periodic structure shown is capable of supporting a propagating wave, then voltage and current at the interface $n$ differ from voltage

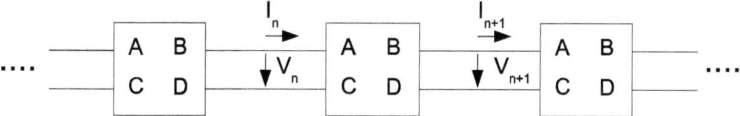

Figure 7.2: Three unit cells of a general 1-D periodic structure

and current at the interface $n+1$ only by a complex factor $e^{\gamma d}$:

$$\begin{bmatrix} V_n \\ I_n \end{bmatrix} = e^{\gamma d} \begin{bmatrix} V_{n+1} \\ I_{n+1} \end{bmatrix} \tag{7.9}$$

where $d$ is the structure period, i.e. the unit cell length, and $\gamma = -\alpha + j\beta$ is the effective propagation constant. Coefficients $\alpha$ and $\beta$ are respectively attenuation- and phase constants of the periodic structure.

From equations 7.8 and 7.9 it is clear that $e^{\pm \gamma d}$ are two independent eigenvalues of the ABCD matrix, describing propagation in both directions along the periodic structure. The associated eigenvectors are $[V_{n+1},\ I_{n+1}]^T$ and $[V_{n+1},\ -I_{n+1}]^T$, respectively.

It practice, periodic structures are not infinite, but consist of a cascade of $N$ unit cells. The ABCD parameters of the cascade arrangement are obtained by taking the $N^{th}$ power of the ABCD parameters of a single unit cell:

$$\begin{bmatrix} A_{cascade} & B_{cascade} \\ C_{cascade} & D_{cascade} \end{bmatrix} = \begin{bmatrix} A & B \\ C & D \end{bmatrix}^N \tag{7.10}$$

The power of a matrix can be computed by its decomposition using eigenvalues $e^{\pm \gamma d}$ and eigenvectors $[V_{n+1},\ \pm I_{n+1}]^T$ (see e.g. [60]):

$$\begin{bmatrix} A & B \\ C & D \end{bmatrix}^N = \begin{bmatrix} V_{n+1} & V_{n+1} \\ I_{n+1} & -I_{n+1} \end{bmatrix} \begin{bmatrix} e^{\gamma d} & 0 \\ 0 & e^{-\gamma d} \end{bmatrix} \begin{bmatrix} V_{n+1} & V_{n+1} \\ I_{n+1} & -I_{n+1} \end{bmatrix}^{-1} \tag{7.11}$$

The eigenvectors $[V_{n+1},\ \pm I_{n+1}]^T$ can be normalized by $\pm I_{n+1}$ to obtain $[\pm Z_b,\ 1]^T$, where $Z_b$ is often referred to as Bloch impedance in the literature and is considered in detail further down. Eq. 7.11 can then be rewritten as:

$$\begin{bmatrix} A & B \\ C & D \end{bmatrix}^N = \begin{bmatrix} Z_b & -Z_b \\ 1 & 1 \end{bmatrix} \begin{bmatrix} e^{\gamma d} & 0 \\ 0 & e^{-\gamma d} \end{bmatrix} \begin{bmatrix} 1 & Z_b \\ -1 & Z_b \end{bmatrix} \frac{1}{2Z_b} \tag{7.12}$$

The result after the matrix multiplication is given below:

$$\begin{bmatrix} A_{cascade} & B_{cascade} \\ C_{cascade} & D_{cascade} \end{bmatrix} = \begin{bmatrix} \cosh(\gamma dN) & Z_b \sinh(\gamma dN) \\ \frac{1}{Z_b}\sinh(\gamma dN) & \cosh(\gamma dN) \end{bmatrix} \quad (7.13)$$

Eq. 7.13 indicates that a periodic structure composed of $N$ identical unit cells behaves as an equivalent transmission line of length $d \cdot N$ with a propagation constant $\gamma$ and a characteristic impedance $Z_b$. Thus, the properties of the overall periodic structure are uniquely determined by its unit cell. It should be noted that both propagation constant and characteristic impedance are highly frequency-dependent. In the following, their properties are considered in detail.

### 7.2.2 Effective Propagation Constant

The effective propagation constant $\gamma$ can be found from the the following eigenvalue equation:

$$\left( \begin{bmatrix} A & B \\ C & D \end{bmatrix} - \begin{bmatrix} e^{\gamma d} & 0 \\ 0 & e^{\gamma d} \end{bmatrix} \right) \cdot \begin{bmatrix} V_{n+1} \\ I_{n+1} \end{bmatrix} = 0 \quad (7.14)$$

The eigenvalues are readily found from the characteristic equation of the matrix:

$$\det \begin{pmatrix} A - e^{\gamma d} & B \\ C & D - e^{\gamma d} \end{pmatrix} = 0$$

which is equivalent to:

$$e^{2\gamma d} - (A+D)e^{\gamma d} + AD - BC = 0 \quad (7.15)$$

Noticing that for a reciprocal network AD-BC=1 and dividing Eq. 7.15 by $e^{\gamma d}$, leads to:

$$\cosh(\gamma d) = \frac{A+D}{2} \quad (7.16)$$

Eq. 7.16 determines the effective propagation constant of a general reciprocal periodic structure. Note that since hyperbolic cosine is an even function, both positive and negative values of the propagation constant $\pm\gamma$ are solutions, indicating propagation of waves in both directions of the periodic structure. Besides, for a travelling wave the absolute value of gamma must have a non-zero imaginary part, i.e. the phase constant $\beta$. Otherwise, the wave is highly attenuated and evanescent.

### 7.2.3 Pass- and Stop-bands

The pass- and stop-band characteristics can be obtained from from Eq. 7.16. To significantly simplify the analysis, the periodic structure is assumed to be lossless. Also, this is often a justified assumption due to very low loss of RFMEMS components and underlying microwave substrates.

In the lossless structure $A$ and $D$ are real numbers, leading to three cases:

1. for $|A + D| \leq 2$, the wave is propagating without attenuation and:

$$\begin{cases} \cos(\beta d) = \dfrac{A+D}{2} \\ \alpha = 0 \end{cases}$$

2. for $A + D > 2$, the wave is evanescent and:

$$\begin{cases} \beta = 0 \\ \cosh(\alpha d) = \dfrac{A+D}{2} \end{cases}$$

3. for $A + D < -2$, the wave is also evanescent and:

$$\begin{cases} \beta = \pi \\ \cosh(\alpha d) = \dfrac{|A+D|}{2} \end{cases}$$

The first case corresponds to a passband and the last two cases describe stopbands in a periodic structure. Thus, the unit cell fully determines frequency dependence of the phase- and attenuation constants. This is used to obtain desired dispersion characteristics for the frequency range of interest. Note that in a lossless periodic structure the attenuation in the stop-band is solely due to reflection.

## 7.2.4 Bloch Impedance

Not only the propagation constant changes with frequency, so does the effective characteristic impedance of a periodic structure, often referred to as the Bloch impedance. In the literature, the term Bloch impedance $Z_b$ implies a characteristic impedance presented to the voltage and current at the reference terminal plane. It can be calculated by rearranging Eq. 7.14:

$$\begin{cases} (A - e^{\gamma d})V_{n+1} = -B \cdot I_{n+1} \\ C \cdot V_{n+1} = -(D - e^{\gamma d})I_{n+1} \end{cases}$$

The Bloch impedance can then be deduced as

$$Z_b = \frac{V_{n+1}}{I_{n+1}} = -\frac{B}{A - e^{\gamma d}} = -\frac{D - e^{\gamma d}}{C} \qquad (7.17)$$

Substituting $e^{\pm \gamma d} = \frac{A+D}{2} \pm \sqrt{\left(\frac{A+D}{2}\right)^2 - 1}$ from Eq. 7.15 into Eq. 7.17 results in:

$$\pm Z_b = -\frac{B}{A - \frac{A+D}{2} \mp \sqrt{\left(\frac{A+D}{2}\right)^2 - 1}}$$

$$= \frac{B}{\frac{D-A}{2} \pm \sqrt{\left(\frac{A+D}{2}\right)^2 - 1}} \qquad (7.18)$$

As will be shown in Sections 8.1, 8.2 and 8.3, the Bloch impedance is strongly frequency-dependent and approaches zero or infinity at the edges of the passband, leading to high reflection and thus, no wave propagation in the stopband.

## 7.2.5 Special Case of Lossless and Symmetrical Unit Cell

As described above, Eq. 7.16 fully describes propagation and attenuation properties of an arbitrary periodic structure with a reciprocal unit cell. Depending on the actual periodic structure, it can possess right- or left-handed properties in different pass-bands, with stop-bands in between.

To greatly simplify the analysis in the following sections, a special case of a lossless and symmetrical unit cell is considered below. Then $A = D$ and $A, D$ are real numbers, hence

Eq. 7.16 simplifies to $\cosh(\gamma d) = A$. The transitions from a pass- to stopband or vice versa take place whenever the magnitude of $A$ crosses unity.

To find phase and attenuation constants, three different cases are considered:

1. if $|A| \leqslant 1$, then

$$\begin{cases} \beta_n d = \pm \arccos(A) + 2\pi n \\ \alpha d = 0 \end{cases}$$

and the wave propagates without attenuation.

2. if $A > 1$, then

$$\begin{cases} \beta_n d = 2\pi n \\ \alpha d = a\cosh(A) \end{cases}$$

and the wave is evanescent.

3. if $A < -1$, then

$$\begin{cases} \beta_n d = \pm\pi + 2\pi n \\ \alpha d = a\cosh(|A|) \end{cases}$$

and the wave is evanescent. Besides, the wave experiences a phase delay of 180 degrees.

For a symmetrical unit cell, the Bloch characteristic impedance given by Eq. 7.18 reduces to:

$$Z_b = \frac{B}{\pm\sqrt{A^2 - 1}} = \pm\sqrt{\frac{B}{C}} \qquad (7.19)$$

# Chapter 8

# Advanced Modelling of RFMEMS Periodic Structures

## 8.1 Lossless Periodic Structure with Shunt Capacitive Loading

### 8.1.1 ABCD-Parameters of a Unit Cell

Periodic structures with shunt capacitive loading are commonly used to design DMTL phase shifters, impedance tuners and other circuits. Two unit cells of a periodic structure with shunt capacitive loading are shown in Fig. 6.1. This section presents a thorough analysis of its properties utilizing the theory of periodic structures.

To facilitate the subsequent analyse, the following parameters of interconnecting unloaded transmission line are defined:

- per unit length capacitance: $C' = \frac{1}{v_{ph,0} Z_0}$
- per unit length inductance: $L' = C' Z_0^2$
- phase constant: $\beta_0 = \frac{\omega}{v_{ph,0}}$
- phase velocity : $v_{ph,0} = \frac{c}{\sqrt{\epsilon_{r,eff}}} = \frac{1}{\sqrt{L'C'}}$

In addition, the MEMS capacitive loading factor is defined as the ratio of the MEMS capacitance $C$ to the transmission line capacitance of the unit cell $C'd$ [61]:

- $LF_C = \frac{C}{C'd}$

ABCD parameters of a unit cell from Fig. 6.1 are derived in the following:

$$\begin{bmatrix} A & B \\ C & D \end{bmatrix} = \begin{bmatrix} \cos(\beta_0 \frac{d}{2}) & jZ_0 \sin(\beta_0 \frac{d}{2}) \\ j\frac{1}{Z_0}\sin(\beta_0 \frac{d}{2}) & \cos(\beta_0 \frac{d}{2}) \end{bmatrix} \begin{bmatrix} 1 & 0 \\ j\omega C & 1 \end{bmatrix} \begin{bmatrix} \cos(\beta_0 \frac{d}{2}) & jZ_0 \sin(\beta_0 \frac{d}{2}) \\ j\frac{1}{Z_0}\sin(\beta_0 \frac{d}{2}) & \cos(\beta_0 \frac{d}{2}) \end{bmatrix} \quad (8.1)$$

or equivalently:

$$\begin{bmatrix} A & B \\ C & D \end{bmatrix} = \begin{bmatrix} \cos(\beta_0 d) - \frac{\omega C Z_0}{2}\sin(\beta_0 d) & j\left(\frac{Z_0^2 \omega C}{2}\cos(\beta_0 d) + Z_0 \sin(\beta_0 d) - \frac{Z_0^2 \omega C}{2}\right) \\ j\left(\frac{\omega C}{2}\cos(\beta_0 d) + \frac{1}{Z_0}\sin(\beta_0 d) + \frac{\omega C}{2}\right) & \cos(\beta_0 d) - \frac{\omega C Z_0}{2}\sin(\beta_0 d) \end{bmatrix} \quad (8.2)$$

In the notations introduced above, the factor $\omega C Z_0$ is equivalent to $\beta_0 v_{ph,0} C Z_0 = \frac{\beta_0 C \sqrt{L'd}}{\sqrt{L'C'}\sqrt{C'd}} = LF_C \beta_0 d$. Thus, the coefficient $A$ depends solely on the phase delay of the unloaded unit cell $\beta_0 d$ and on the loading factor $LF_C$. Finally, the Eq. 8.2 can be rewritten as:

$$\begin{cases} A = \cos(\beta_0 d) - \beta_0 d \frac{LF_C}{2}\sin(\beta_0 d) \\ B = j\left(\beta_0 d \frac{Z_0 LF_C}{2}\cos(\beta_0 d) + Z_0 \sin(\beta_0 d) - \beta_0 d \frac{Z_0 LF_C}{2}\right) \\ C = j\left(\beta_0 d \frac{LF_C}{2Z_0}\cos(\beta_0 d) + \frac{1}{Z_0}\sin(\beta_0 d) + \beta_0 d \frac{LF_C}{2Z_0}\right) \\ D = \cos(\beta_0 d) - \beta_0 d \frac{LF_C}{2}\sin(\beta_0 d) \end{cases} \quad (8.3)$$

### 8.1.2 Effective Propagation Constant

As it was shown, frequency-dependent effective propagation constant can be obtained from the ABCD analysis described in Section 7.2.2. Besides, whenever the magnitude of $A$ crosses unity, pass- and stop-bands appear.

In the following, three values of loading factors $LF_C$ are considered:

1. $LF_C$=0, i.e. no MEMS

2. $LF_C$=0.5, i.e. small MEMS loading

3. $LF_C$=1, i.e. high MEMS loading

Fig. 8.1 shows values of $A$ versus phase delay of a single unloaded unit cell $\beta_0 d$ calculated using Eq. 8.3 for three loading factors. As expected, for the unloaded case ($LF_C = 0$), $A$ is defined by a pure cosine function, whose magnitude never exceeds unity and no pass- and stop-bands are expected. As MEMS loading is introduced, for some frequency ranges the magnitude of $A$ exceeds unity and pass- and stop-bands appear.

Phase delay and attenuation of a loaded unit cell can be calculated from $A$ using Eq. 7.16. It should be stressed that since the inverse cosine is a multi-valued function, care should be taken to select the correct values of the phase delay $\beta d$. Since simply taking the principal value of the inverse cosine function would lead to incorrect results as shown in Fig. 8.2, sometimes reported in the literature. With the principal value calculation, both unloaded and loaded lines appear to show left-handed behaviour in half of the passbands, which is unphysical for this type of periodic structure.

The calculated phase delay $\beta d$ and attenuation $\alpha d$ versus phase delay of the unloaded unit cell $\beta_0 d$ are shown in Fig. 8.1. In the notations of Section 7.1, $\beta d$ refers to the phase delay of the fundamental space harmonics, which also gives the phase delay of the overall wave. At low frequencies, the structure supports a propagating "slow" wave, since its phase delay $\beta d$ is larger than $\beta_0 d$ - that of the unloaded transmission line.

As $\beta d$ reaches $\pi$ (or any multiple of $\pi$), the wave stops to propagate and the group velocity, given by the slope of $\beta d$, is zero. This is explained by an effect similar to Bragg reflection in optics, when the propagating and reflected waves form a standing wave, as was also derived from the space harmonic analysis in Section 7.1. The condition $\beta d = m\pi$ corresponds to the so-called Bragg cut-off frequencies, at which the incoming wave is reflected back. Above these cut-off frequencies, there are stopbands with highly-attenuated or evanescent waves. Since the considered structure is lossless, the attenuation is due to wave reflection (similar to filters). From Fig. 8.1 one notices that stopbands end whenever $\beta d = \beta_0 d$ for $\beta_0 d = m\pi$ for all loading factors. This indicates that the standing wave has its zeros at the locations of the discontinuities and thus, capacitive loading has no effect on its propagation characteristics.

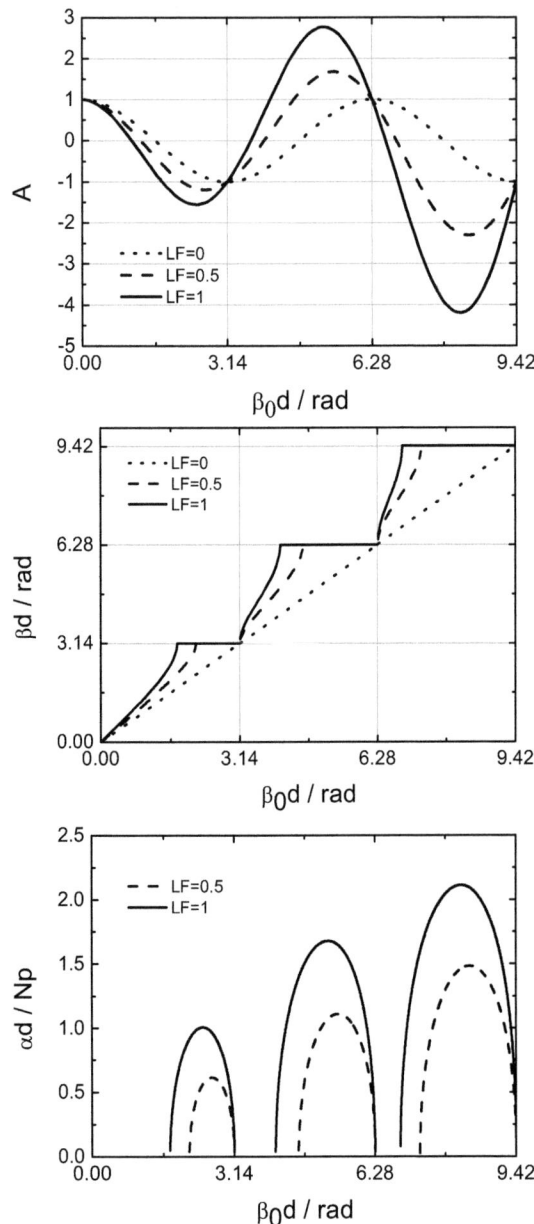

Figure 8.1: Calculated coefficient $A$, phase delay $\beta d$ and attenuation $\alpha d$ of a capacitively-loaded unit cell versus phase delay of an unloaded unit cell $\beta_0 d$

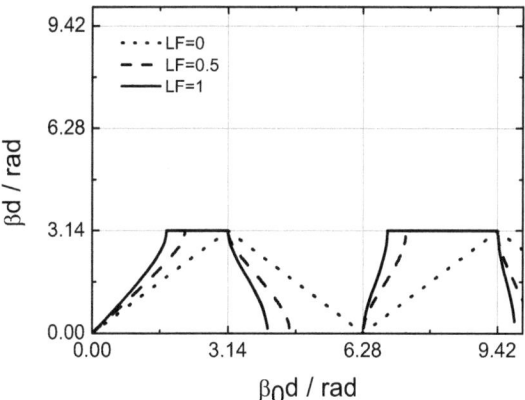

Figure 8.2: Sometimes reported in the literature: phase delay erroneously calculated using principal values of inverse cosine, leading to unphysical results

### 8.1.3 Bloch Impedance

The Bloch impedance for this periodic structure was calculated using Eq. 7.19 assuming a 50 $\Omega$ characteristic impedance of the unloaded transmission line. The result is shown in Fig. 8.3. At low frequencies, the impedance is smaller than 50 $\Omega$ due to capacitive loading. As the frequency increases, the impedance starts to decrease rapidly reaching zero at the first passband cut-off and also at the edges of all odd passbands. At the edges of the all even passbands, the impedance approaches infinity. In the stopbands, the Bloch impedance is purely imaginary (not shown in Fig. 8.3.). This clearly indicates that this strong frequency dependence of the Bloch impedance must be considered in the circuit design and that the conventional low-frequency model strongly deviates from the reality.

### 8.1.4 Cut-off Frequencies of Pass- and Stopbands

In the following, accurate equations for the cut-off frequencies of pass- and stopbands are derived. Cut-off frequencies where stopbands end ($f_{c\_stop,m}$) are independent of the MEMS loading and are found from the condition $\beta_0 d = m\pi$ (note for future calculations, that this is equivalent to $\sin(\beta_0 d) = 0$), leading to:

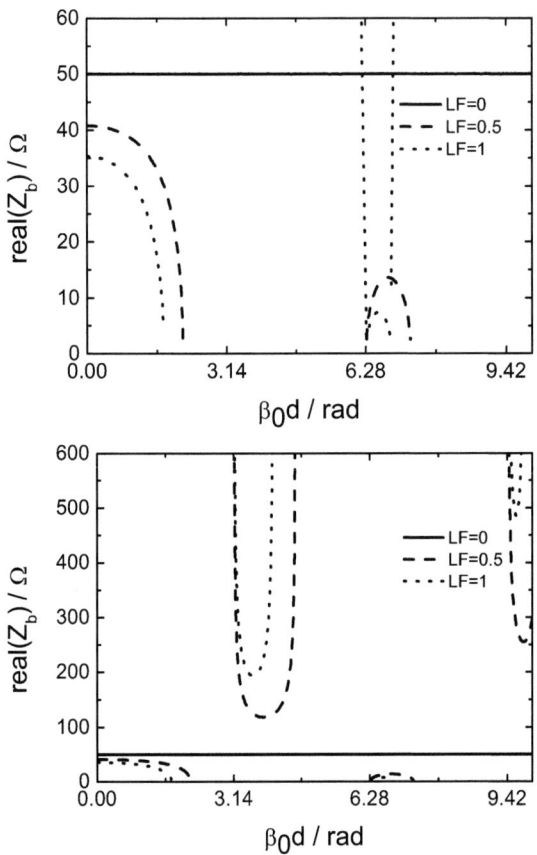

Figure 8.3: Bloch impedance of a capacitively-loaded 50 $\Omega$ transmission line versus phase delay of an unloaded unit cell $\beta_0 d$

$$f_{c\_stop,m} = \frac{m}{2d\sqrt{L'C'}}$$

However, the cut-off frequencies where passbands end ($f_{c\_pass,m}$) are dependent on the loading factor and are not so readily found. As it was shown in Section 7.2.5, transitions between pass- and stopbands take place whenever the magnitude of $A$ crosses unity. Thus, the passband cut-off frequencies must be obtained from the two conditions:

$$\cos(\beta_0 d) - \beta_0 d \frac{LF_C}{2} \sin(\beta_0 d) = 1 \tag{8.4}$$

$$\cos(\beta_0 d) - \beta_0 d \frac{LF_C}{2} \sin(\beta_0 d) = -1 \tag{8.5}$$

It was shown above, that $\sin(\beta_0 d) = 0$ at all $f_{c\_stop,m}$. Thus, $\sin(\beta_0 d) \neq 0$ at all $f_{c\_pass,m}$ and Equations 8.4 and 8.5 can be divided by $\sin(\beta_0 d)$ to obtain two cases:

$$-LF_C \frac{\beta_0 d}{2} = \tan(\frac{\beta_0 d}{2}) \tag{8.6}$$

and

$$LF_C \frac{\beta_0 d}{2} = \cot(\frac{\beta_0 d}{2}) \tag{8.7}$$

Equations 8.6 and 8.7 are transcendental and can not be solved analytically. However, within this work accurate approximate solutions to these equations were obtained. In this section only final results are presented, while the derivations are given in Appendices A and B. The derived approximate solutions of Equations 8.6 and 8.7 are given by Equations 8.8 and 8.9 respectively:

$$\begin{cases} \beta_0 d \approx -2\dfrac{[\arctan(a) - m\pi](1+a^2) - a}{1 + a^2 + LF_C} \\ a = (2m-1)\dfrac{\pi}{2} \end{cases} \tag{8.8}$$

$$\begin{cases} \beta_0 d \approx 2\dfrac{[\arctan(a) - m\pi](1+a^2) + a}{1 + a^2 + LF_C} \\ a = (2m-1)\dfrac{\pi}{4} \end{cases} \tag{8.9}$$

Excellent accuracy is obtained with two iterations of the proposed algorithm (see Appendices A and B). After two iterations, the following cut-off conditions of the first four passbands were calculated for the example loading factors $LF_C$ of 0.1, 1 and 3: Once $\beta_0 d_{c\_pass,m}$

Table 8.1: Caption

| $LF_C$ | $\beta_0 d_{c\_pass,1}$ | $\beta_0 d_{c\_stop,1}$ | $\beta_0 d_{c\_pass,2}$ | $\beta_0 d_{c\_stop,2}$ | $\beta_0 d_{c\_pass,3}$ | $\beta_0 d_{c\_stop,3}$ | $\beta_0 d_{c\_pass,4}$ |
|---|---|---|---|---|---|---|---|
| 0.1 | 2.86 | 3.14 | 5.73 | 6.28 | 8.61 | 9.42 | 11.52 |
| 1 | 1.72 | 3.14 | 4.06 | 6.28 | 6.85 | 9.42 | 9.83 |
| 3 | 1.07 | 3.14 | 3.52 | 6.28 | 6.49 | 9.42 | 9.56 |

are found, the corresponding cut-off frequencies are readily calculated:

$$f_{c\_pass,m} = \frac{\beta_0 d_{c\_pass,m}}{2\pi d \sqrt{L'C'}}$$

As table 8.1, an increased loading factor reduces cut-off frequencies of the passbands and makes the passbands narrower. Besides, for a specific loading factor higher passbands are always narrower as the lower ones. This is also clear from Fig. 8.1.

In the conventional RFMEMS periodic structures modelling, cut-off frequencies of the first passbands would be calculated with Eq. 6.2. For the loading factor considered here, the results would be:

- for $LF_C$=0.1: $\beta_0 d_{c\_pass,1} = 1.9$
- for $LF_C$=1: $\beta_0 d_{c\_pass,1} = 1.41$
- for $LF_C$=3: $\beta_0 d_{c\_pass,1} = 1$

Comparing these values to the ones in table 8.1, it is clear, that Eq. 6.2 leads to incorrect results, especially for small loading factors often used in phase shifters.

### 8.1.5 Experimental Validation of the Model

To validate the developed model, a periodic structure composed of a coplanar line loaded with RFMEMS switchable capacitors from Section 4.4 was designed. As the SEM image in

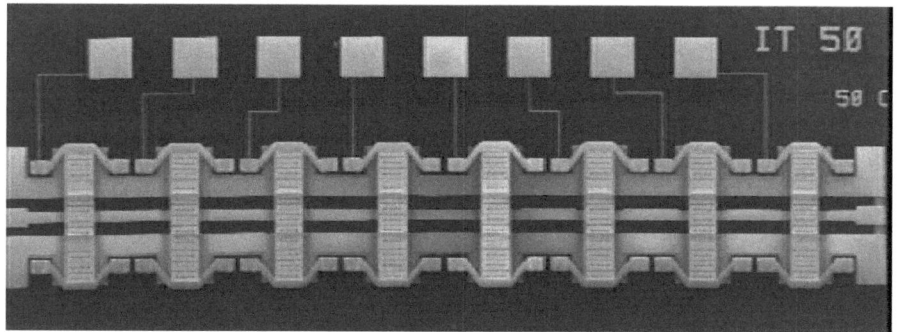

Figure 8.4: SEM image of periodic structure in ISiT technology composed of a coplanar line loaded with RFMEMS capacitors presented in Section 4.4. The circuit operates as an impedance tuner and the results are presented in Section 10.3.

Fig. 8.4 shows, the periodic structure consists of eight unit cells, each formed by a section of transmission line and one switchable capacitor. To explore the effect of different loading factors on electromagnetic properties of the periodic structure, two different bias voltage configurations were studied: 1) with all MEMS capacitors in the up-state, i.e. without bias voltage and 2) with all MEMS capacitors in the down-state, i.e. with 40 V applied to the CPW centre conductor and to the MEMS bridges. Using the capacitive loading factor definition given in Section 8.1.1, geometrical dimensions of the transmission line and MEMS capacitance values, $LF_C$ are calculated to be 0.54 and 1.18 for the two states respectively.

S-parameters of the periodic structure measured up to 110 GHz for both bias configurations are shown in Fig 8.5. In contrast to the unbiased "all MEMS up" state, the insertion- and return losses in the "all MEMS down" state sharply increase at higher frequencies. This indicates a cut-off of the first passband, i.e. a beginning of the first stopband at around 90 GHz. Higher stop- and passbands are beyond the measurement frequency limit. As expected, the phase delay is larger in the down-state due to a larger loading factor.

By converting the S-parameters into ABCD-parameters and applying the equations described in Section 7.2.2, the frequency-dependent propagation constant was calculated. The results are plotted in Fig. 8.6 together with theoretical predictions of the model for the corresponding loading factors. In general, the results agree quite well. The differences between measured and theoretical curves can be explained by the fact that the structure

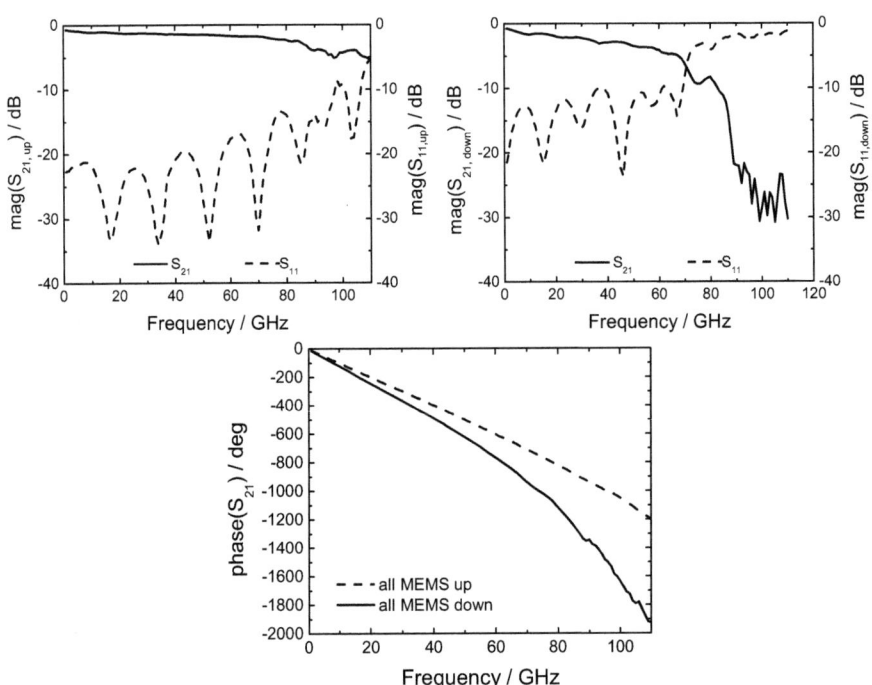

Figure 8.5: Measured S-parameters of the periodic structure shown in Fig. 8.4. Measurements are taken for two configurations: all MEMS up (left) and all MEMS down (right). Actuation voltage is 40 V. The structure is approaching its first stopband at about 90 GHz.

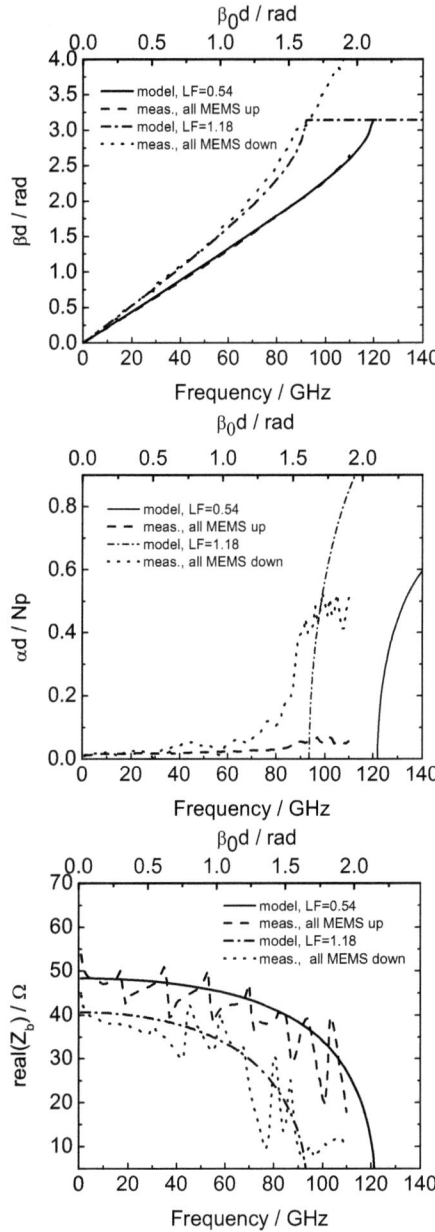

Figure 8.6: Measured and modelled phase delay, attenuation and Bloch impedance of the periodic structure shown in Fig. 8.4.

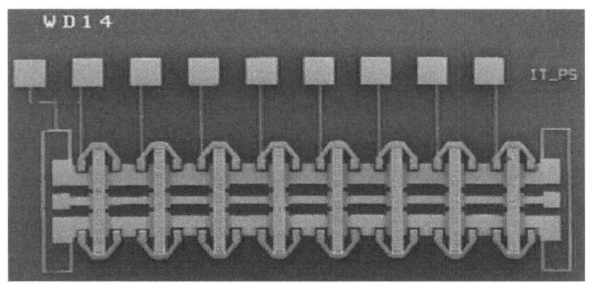

Figure 8.7: SEM image of periodic structure in ISiT technology composed of a coplanar line loaded with RFMEMS capacitors presented in Section 4.5. The circuit operates as an impedance tuner and the results are presented in Section 10.2.

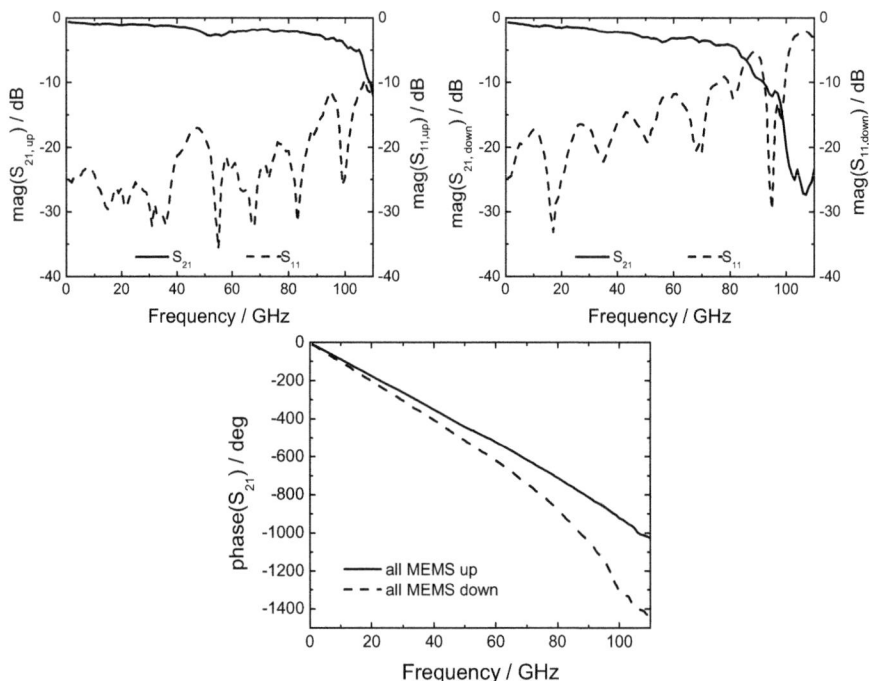

Figure 8.8: Measured S-parameters of the periodic structure shown in Fig. 8.7. Measurements are taken for two configurations: all MEMS up (left) and all MEMS down (right). Actuation voltage is 40 V. The structure is approaching its first stopband at about 90 GHz.

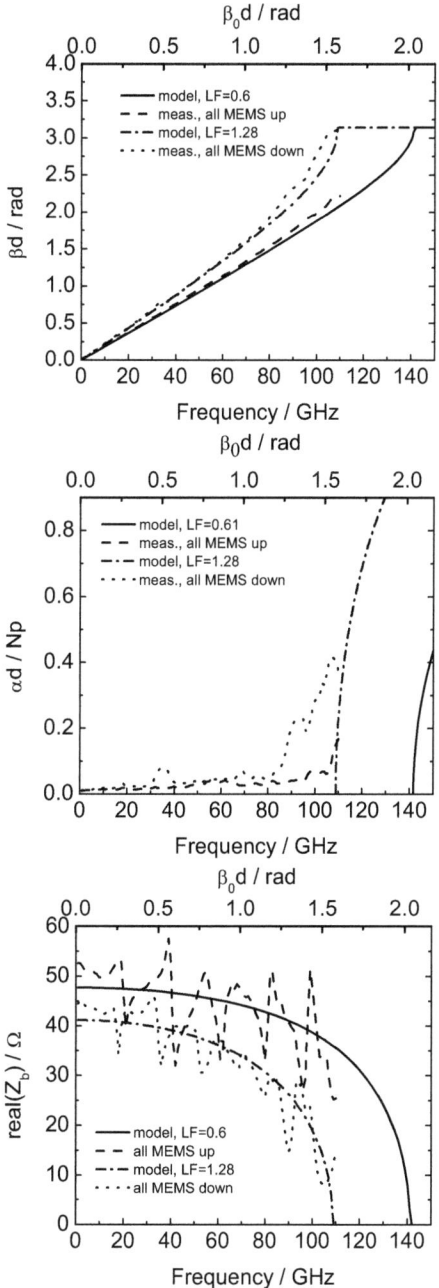

Figure 8.9: Measured and modelled phase delay, attenuation and Bloch impedance of the periodic structure shown in Fig. 8.7.

considered in Fig. 8.1 is assumed to be lossless. As a result it possesses a purely imaginary and a purely real propagation constant within pass- or stopbands respectively, i. e. no attenuation in the passband and a constant phase delay in the stopband. The measured structure, however, has some small losses. It can be shown [59] that the propagation constant of a lossy structure will be complex in both stop- and pass regions. This is in agreement with the measured results, showing some attenuation in the passband and an increasing phase delay in the stopband. An important observation is that the measured attenuation constant greatly increases shortly before the cut-off frequency. Thus, this frequency region should be avoided in circuit operation and an accurate estimation of cut-off frequencies is essential. As a rule of thumb, low-loss performance is maintained up to about 70 % of the cut-off frequency of the first passband

The Bloch impedance was calculated applying equations described in Section 7.2.4 and is shown in Fig. 8.6. There is a good agreement between the measurement and the model, despite some fluctuations of the measured impedance value. The fluctuations are due to the fact that the structure is not terminated with $Z_b$, creating periodic modulation in the reflection coefficient, having a large influence on the extracted Bloch impedance values.

Using the approximate solutions derived in Appendices A and B, the cut-off of the first passband for $LF_C = 0.54$ takes place at $\beta_0 d_{c\_pass,1}$=2.1 or 119 GHz. As Fig. 8.6 shows, this is beyond the measurement limit. For the down-state and the corresponding loading factor $LF_C = 1.18$, the cut-off condition is obtained from $\beta_0 d_{c\_pass,1}$=1.61 or 91 GHz and agrees well with the measured value. This confirms the accuracy of the approximate equations for the cut-off frequencies derived in Appendices A and B.

A similar analysis was conducted with another periodic structure, loaded with RFMEMS switchable capacitors presented in Section 4.5. The SEM image of the structure is shown in Fig. 8.7. The measured S-parameters in the up- and down states can be seen in Fig. 8.8. In the up-state, the loading factor is equal to 0.61, whereas in the down-state it is 1.28. The extracted propagation constant and impedance together with the theoretically predicted values are plotted in Fig. 8.9. Similar conclusions can be made as for the previous capacitive periodic structure. This time, the first cut-off frequency is around 110 GHz and there is a good agreement between measured and calculated values. The cut-off condition predicted by the model for $LF_C = 0.61$ is $\beta_0 d_{c\_pass,1}$=2.03 or 160 GHz and is beyond the measure-

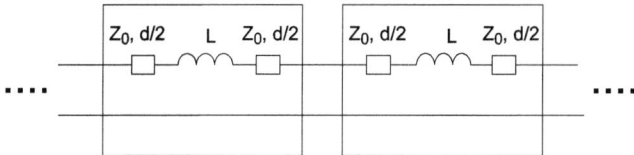

Figure 8.10: Transmission line with characteristic impedance $Z_0$ loaded with series inductors $L$ after each period $d$

ment limit. In the down-state, the predicted cut-off is at $\beta_0 d_{c\_pass,1}$=1.56 or 109 GHz and corresponds well to the experimental value.

## 8.2 Lossless Periodic Structure with Series Inductive Loading

### 8.2.1 ABCD-Parameters of a Unit Cell

RFMEMS periodic structures with inductive loading are rarely reported in the literature, since switchable RFMEMS inductors are not as widely available as RFMEMS capacitors. On the other hand, for some applications it may be advantageous to utilize inductive loading, which is studied in detail in the following.

Similar to the capacitive loading case, the inductive loading factor is defined as the ratio of the MEMS inductance to the unit cell inductance of the unloaded transmission line:

- $LF_L = \frac{L}{L'd}$

Two unit cells of an inductively-loaded periodic structure are shown in Fig. 8.10. The ABCD parameters of a unit cell are derived in the following:

$$\begin{bmatrix} A & B \\ C & D \end{bmatrix} = \begin{bmatrix} \cos(\beta_0 \frac{d}{2}) & jZ_0 \sin(\beta_0 \frac{d}{2}) \\ j\frac{1}{Z_0}\sin(\beta_0 \frac{d}{2}) & \cos(\beta_0 \frac{d}{2}) \end{bmatrix} \begin{bmatrix} 1 & j\omega L \\ 0 & 1 \end{bmatrix} \begin{bmatrix} \cos(\beta_0 \frac{d}{2}) & jZ_0 \sin(\beta_0 \frac{d}{2}) \\ j\frac{1}{Z_0}\sin(\beta_0 \frac{d}{2}) & \cos(\beta_0 \frac{d}{2}) \end{bmatrix}$$

(8.10)

$$= \begin{bmatrix} \cos(\beta_0 d) - \frac{\omega L}{2Z_0}\sin(\beta_0 d) & j\left(\frac{\omega L}{2}\cos(\beta_0 d) + Z_0 \sin(\beta_0 d) + \frac{\omega L}{2}\right) \\ j\left(\frac{\omega L}{2Z_0^2}\cos(\beta_0 d) + \frac{1}{Z_0}\sin(\beta_0 d) - \frac{\omega L}{2Z_0^2}\right) & \cos(\beta_0 d) - \frac{\omega L}{2Z_0}\sin(\beta_0 d) \end{bmatrix}$$

The factor $\frac{\omega L}{Z_0}$ is equivalent to $\frac{\beta_0 V_{ph,0} L}{Z_0} = \frac{\beta_0 L\sqrt{C'd}}{\sqrt{L'C'}\sqrt{L'd}} = LF_L \beta_0 d$. Then Eq.8.11 can be rewritten as:

$$\begin{cases} A = \cos(\beta_0 d) - \beta_0 d \frac{LF_L}{2} \sin(\beta_0 d) \\ B = j\left(\beta_0 d \frac{Z_0 LF_L}{2} \cos(\beta_0 d) + Z_0 \sin(\beta_0 d) + \beta_0 d \frac{Z_0 LF_L}{2}\right) \\ C = j\left(\beta_0 d \frac{LF_L}{2Z_0} \cos(\beta_0 d) + \frac{1}{Z_0} \sin(\beta_0 d) - \beta_0 d \frac{LF_L}{2Z_0}\right) \\ D = \cos(\beta_0 d) - \beta_0 d \frac{LF_L}{2} \sin(\beta_0 d) \end{cases}$$

(8.11)

### 8.2.2 Effective Propagation Constant

From Eq. 8.11 it is clear that the coefficient $A$ is the same as in case of the capacitive loading, with the only difference that the inductive loading factor should be used instead. Consequently, the attenuation $\alpha d$ and phase delay $\beta d$ are identical to that of the capacitive loading case shown in Fig. 8.1.

### 8.2.3 Bloch Impedance

Contrary to the propagation constant, the Bloch impedance is very different to that in the shunt capacitive loading case. As before, the impedance was calculated using Eq. 7.19 assuming a 50 Ω characteristic impedance of the unloaded transmission line. The result can be seen in Fig. 8.11. At low frequencies, the impedance is higher than 50 Ω due to the inductive loading. As frequency increases, the impedance value raises, reaching infinity at the cut-off of the first passband. At the edges of all even passbands, the impedance equals zero. This is also in agreement with results reported in [62]. Generally, one can conclude that the variation of the Bloch impedance versus frequency is complementary to that in case of the capacitive loading, shown in Fig. 8.3

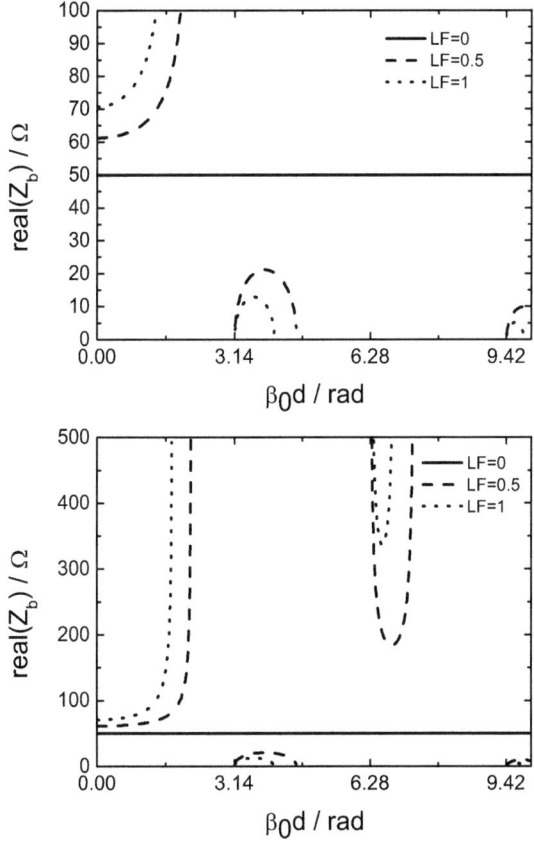

Figure 8.11: Bloch impedance of an inductively-loaded 50 $\Omega$ transmission line versus phase delay of an unloaded unit cell $\beta_0 d$

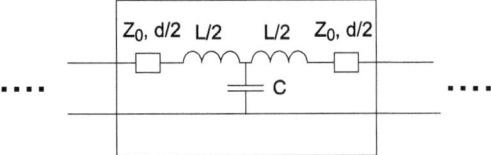

Figure 8.12: Unit cell of an inductively and capacitively-loaded periodic structure

### 8.2.4 Cut-off Frequencies of Pass- and Stopbands

As explained above, transitions between pass- and stopbands occur whenever the magnitude of $A$ crosses unity. Since $A$ is identical to that in the capacitive loading case, cut-off frequencies of series inductive loading are the same too (see Section 8.1.4).

## 8.3 Lossless Periodic Structure with Series Inductive and Shunt Capacitive Loading

### 8.3.1 ABCD-Parameters of a Unit Cell

As was shown in previous sections, in case of shunt capacitive or series inductive loading the Bloch impedance strongly depends on the loading factor. In other words, it is not possible to obtain the same Bloch impedance for different loading. As a consequence, circuits like DMTL phase shifters suffer from inherent impedance mismatch, since they vary the phase velocity by modifying the loading factor. In addition, strong frequency dispersion of the impedance limits the bandwidth of circuits utilizing such periodic structures.

To overcome these problems, a periodic structure combining shunt capacitive and series inductive loading is suggested in this section. One unit cell of such a structure is shown in Fig. 8.12. ABCD parameters of a unit cell are written below:

$$\begin{bmatrix} A & B \\ C & D \end{bmatrix} = \begin{bmatrix} \cos(\beta_0 \frac{d}{2}) & jZ_0 \sin(\beta_0 \frac{d}{2}) \\ j\frac{1}{Z_0} \sin(\beta_0 \frac{d}{2}) & \cos(\beta_0 \frac{d}{2}) \end{bmatrix} \begin{bmatrix} 1 & j\omega\frac{L}{2} \\ 0 & 1 \end{bmatrix} \begin{bmatrix} 1 & 0 \\ j\omega C & 1 \end{bmatrix} \begin{bmatrix} 1 & j\omega\frac{L}{2} \\ 0 & 1 \end{bmatrix} \times \begin{bmatrix} \cos(\beta_0 \frac{d}{2}) & jZ_0 \sin(\beta_0 \frac{d}{2}) \\ j\frac{1}{Z_0} \sin(\beta_0 \frac{d}{2}) & \cos(\beta_0 \frac{d}{2}) \end{bmatrix} \quad (8.12)$$

or equivalently:

$$\begin{cases} A = \left(1 - \dfrac{\omega^2 LC}{2}\right)\cos(\beta_0 d) - \left[\omega C Z_0 + \left(2 - \dfrac{\omega^2 LC}{2}\right)\dfrac{\omega L}{2Z_0}\right]\dfrac{\sin(\beta_0 d)}{2} \\ B = j\left[\dfrac{\omega C Z_0^2}{2} + \left(2 - \dfrac{\omega^2 LC}{2}\right)\dfrac{\omega L}{4}\right]\cos(\beta_0 d) + j\left(1 - \dfrac{\omega^2 LC}{2}\right)Z_0 \sin(\beta_0 d) - \\ \quad - \dfrac{\omega C Z_0^2}{2} + \left(2 - \dfrac{\omega^2 LC}{2}\right)\dfrac{\omega L}{4} \\ C = j\left[\dfrac{\omega C}{2} + \left(2 - \dfrac{\omega^2 LC}{2}\right)\dfrac{\omega L}{4Z_0^2}\right]\cos(\beta_0 d) + j\left(1 - \dfrac{\omega^2 LC}{2}\right)\dfrac{1}{Z_0}\sin(\beta_0 d) + \\ \quad + \dfrac{\omega C}{2} - \left(2 - \dfrac{\omega^2 LC}{2}\right)\dfrac{\omega L}{4Z_0^2} \\ D = \left(1 - \dfrac{\omega^2 LC}{2}\right)\cos(\beta_0 d) - \left[\omega C Z_0 + \left(2 - \dfrac{\omega^2 LC}{2}\right)\dfrac{\omega L}{2Z_0}\right]\dfrac{\sin(\beta_0 d)}{2} \end{cases} \quad (8.13)$$

Using the same capacitive and inductive loading factor notations as in Section 8.1 and 8.2, results in the following equivalents:

- $\omega C Z_0 = LF_C \beta_0 d$

- $\dfrac{\omega L}{Z_0} = LF_L \beta_0 d$

- $\omega^2 LC = LF_C LF_L (\beta_0 d)^2$

Then, Eq. 8.13 can be rewritten as follows:

$$\begin{cases} A = \left(1 - \dfrac{LF_C LF_L(\beta_0 d)^2}{2}\right)\cos(\beta_0 d) - \left[LF_C \beta_0 d + \left(2 - \dfrac{LF_C LF_L(\beta_0 d)^2}{2}\right)\dfrac{LF_L \beta_0 d}{2}\right]\dfrac{\sin(\beta_0 d)}{2} \\ B = j\left[\dfrac{LF_C Z_0 \beta_0 d}{2} + \left(2 - \dfrac{LF_C LF_L(\beta_0 d)^2}{2}\right)\dfrac{LF_L Z_0 \beta_0 d}{4}\right]\cos(\beta_0 d) + \\ \quad + j\left(1 - \dfrac{LF_C LF_L(\beta_0 d)^2}{2}\right)Z_0 \sin(\beta_0 d) - j\dfrac{LF_C Z_0 \beta_0 d}{2} + j\left(2 - \dfrac{LF_C LF_L(\beta_0 d)^2}{2}\right)\dfrac{LF_L Z_0 \beta_0 d}{4} \\ C = j\left[\dfrac{LF_C \beta_0 d}{2Z_0} + \left(2 - \dfrac{LF_C LF_L(\beta_0 d)^2}{2}\right)\dfrac{LF_L \beta_0 d}{4Z_0}\right]\cos(\beta_0 d) + \\ \quad + j\left(1 - \dfrac{LF_C LF_L(\beta_0 d)^2}{2}\right)\dfrac{1}{Z_0}\sin(\beta_0 d) + j\dfrac{LF_C \beta_0 d}{2Z_0} - j\left(2 - \dfrac{LF_C LF_L(\beta_0 d)^2}{2}\right)\dfrac{LF_L \beta_0 d}{4Z_0} \\ D = \left(1 - \dfrac{LF_C LF_L(\beta_0 d)^2}{2}\right)\cos(\beta_0 d) - \left[LF_C \beta_0 d + \left(2 - \dfrac{LF_C LF_L(\beta_0 d)^2}{2}\right)\dfrac{LF_L \beta_0 d}{2}\right]\dfrac{\sin(\beta_0 d)}{2} \end{cases}$$

(8.14)

## 8.3.2 Bloch Impedance

The motivation behind using the periodic structure with both inductive and capacitive loading is to obtain a Bloch impedance which is much less dependent on the loading factor. It is intuitively expected that capacitive and inductive loadings should exactly balance each other, i.e. $LF_C$ must be equal to $LF_L$. It also follows using Eq. 7.19 and the ABCD-parameters derived above given by Eq. 8.14. Using these two equations, it can be easily shown that the condition $Z_b = Z_0$ is equivalent to:

$$\frac{LF_C \beta_0 d}{2} = \left(2 - \frac{LF_C LF_L (\beta_0 d)^2}{2}\right) \frac{LF_L \beta_0 d}{4} \tag{8.15}$$

For low frequencies, $\beta_0 d$ is small and the second term in the curly brackets can be neglected. Then it is clear that $LF_C$ must be equal to $LF_L$ for the equality to be true, i.e. for $Z_b = Z_0$. Thus, by using the same loading factors for inductive and capacitive loadings, the Bloch impedance at low frequencies is independent of the loading factor and is equal to the unloaded line impedance $Z_0$. Although the condition $Z_b = Z_0$ is true only for very low frequencies or low loading factors, the Bloch impedance stays close to the unloaded line impedance for a relatively large bandwidth. This is the key advantage of this periodic structure for DMTL phase shifters, since it completely resolves the trade-off between the phase shift and mismatch.

The calculated Bloch impedance is shown in Fig. 8.13 ($Z_0 = 50\ \Omega$). As compared to impedances in case of shunt capacitive or series inductive loading shown in Figures 8.11 and 8.13, the results are quite different. At low frequencies, the impedances are the same for all loading factors and equal to the unloaded line impedance. This is true provided that the capacitive and inductive loading factors are equal. As frequency increases, the impedance deviates from the initial value. However the deviation is significantly smaller, as in case of only shunt capacitive or series inductive loading. In the second passband, the impedance is infinite at the edges, but is quite close to the unloaded line impedance in the band centre. Thus, a very good impedance match is possible for different loading factors and over much larger bandwidth.

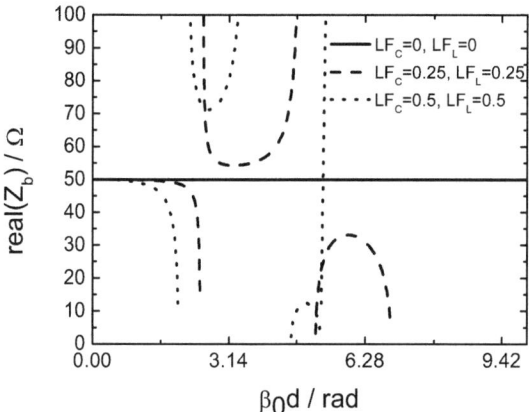

Figure 8.13: Bloch impedance of 50 Ω transmission line loaded with series inductors and shunt capacitors versus phase delay of an unloaded unit cell $\beta_0 d$

### 8.3.3 Effective Propagation Constant

The condition of equal capacitive and inductive loading is used in the following. As in the previous analysis, the effective propagation constant can be found from the coefficient $A$. Fig. 8.14 shows calculated $A$, phase delay $\beta d$ and attenuation constant $\alpha d$ versus phase delay of an unloaded unit cell $\beta_0 d$. To obtain similar overall loading as considered before for shunt capacitive and series inductive cases, inductive and capacitive loading factors are reduced by two, i.e. $LF_C = LF_L = \frac{LF}{2}$. Three sets of loading factors are considered. Comparing Fig. 8.14 to Fig. 8.1, one can see that as in case of shunt capacitive and series inductive loading, there is a slow travelling wave in the passbands and high attenuation in the stopbands. However, the first stopband is much more narrow, especially for small loading. Besides, cut-off frequencies of all stopbands differ for different loading factors, which was not the case for either solely shunt capacitive or series inductive loading.

It is mentioned above that reducing the loading factors by two results in approximately the same loading. A more exact analysis reveals, that in case of combined inductive and capacitive loading with $LF_C = LF_L = \frac{LF}{2}$, the phase velocity is lower as compared to either purely capacitive or purely inductive loading with $LF_C = LF_L = LF$. This is shown below by considering low-frequency approximations of phase velocities for the three cases:

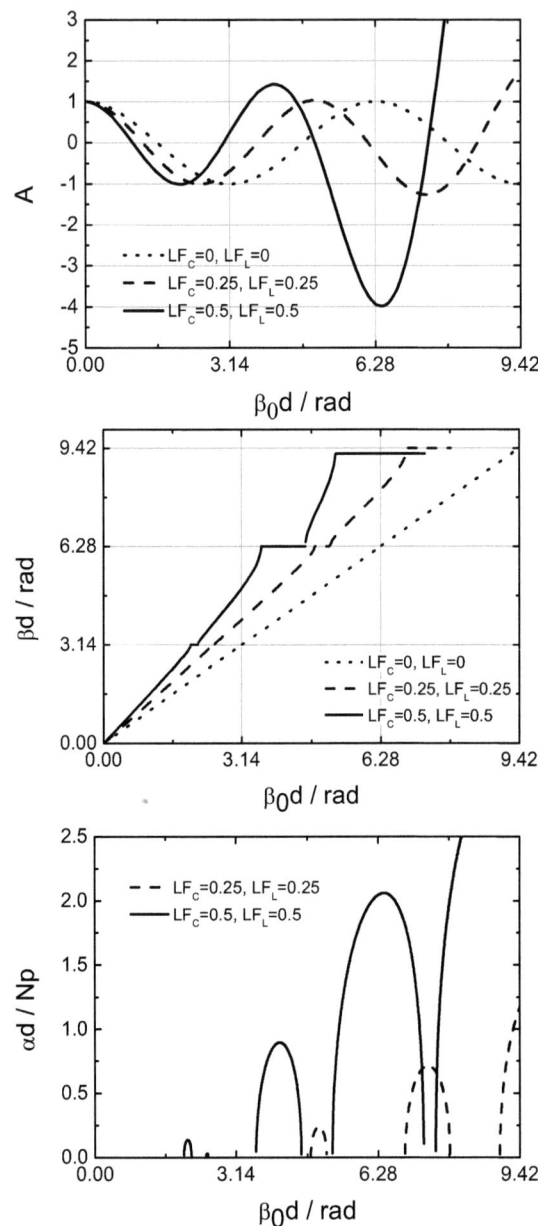

Figure 8.14: Calculated coefficient $A$, phase delay $\beta d$ and attenuation $\alpha d$ of a capacitively- and inductively loaded unit cell versus phase delay of an unloaded unit cell $\beta_0 d$

Figure 8.15: Micrograph of a periodic structure in FBK technology composed of a coplanar line loaded with RFMEMS shunt capacitors presented in Section 5.4 and series inductors presented in Section 5.3.

- capacitive loading: $V_{ph,C} = \frac{1}{\sqrt{L'\left(C'+\frac{C}{d}\right)}} = \frac{V_{ph,0}}{\sqrt{1+LF}}$

- inductive loading: $V_{ph,L} = \frac{1}{\sqrt{C'\left(L'+\frac{L}{d}\right)}} = \frac{V_{ph,0}}{\sqrt{1+LF}}$

- capacitive and inductive loading: $V_{ph,LC} = \frac{1}{\sqrt{\left(C'+\frac{C}{2d}\right)\left(L'+\frac{L}{2d}\right)}} = \frac{V_{ph,0}}{1+\frac{LF}{2}}$

Since $1 + \frac{LF}{2}$ is larger than $\sqrt{1+LF}$, the phase velocity for combined inductive and capacitive loading is lower, i.e. the phase shift is larger, which is very advantageous for phase shifters.

### 8.3.4 Cut-off Frequencies of Pass- and Stopbands

As before, cut-off frequencies of pass- and stopbands can be found from the condition that the absolute value of the coefficient $A$ crosses unity. However, due to much more involved mathematics, simple closed-form equations can not be derived and cut-off frequencies should be found by graphical methods, i.e. by plotting coefficient A versus frequency and equating its absolute value to unity.

### 8.3.5 Experimental Validation of the Model

The developed model was validated with measurements of a periodic structure made of a coplanar line loaded with shunt RFMEMS capacitors from Section 5.4 and series RFMEMS inductors from Section 5.3. A micrograph of the fabricated periodic structure can be seen in Fig. 8.15. To make the structure more compact, its unit cell is composed of a single series inductor and a shunt capacitor, in contrast to the symmetrical unit cell with two series

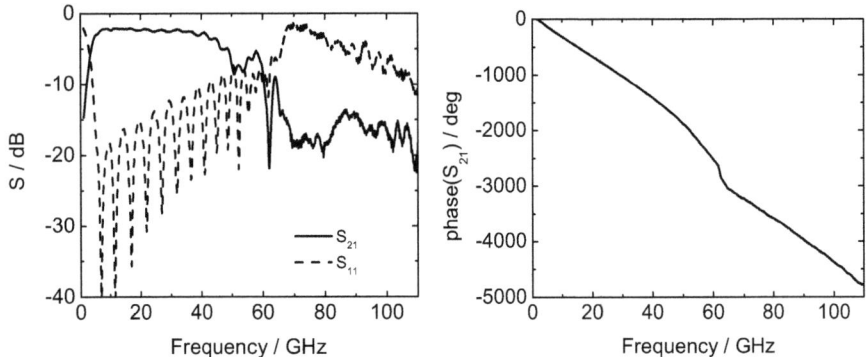

Figure 8.16: Measured S-parameters of the periodic structure shown in Fig. 8.15. The first stopband appears around 80 GHz.

inductors as shown in Fig. 8.12. The lengths of capacitive and inductive parts of the unit cell are $d_c$ = 195 µm and $d_l$ = 221 µm respectively.

Measured S-parameters of the structure up to 110 GHz are shown in Fig. 8.16. Around 70-80 GHz the return- and insertion losses sharply increase, indicating a stopband at this frequencies. Extracted propagation constant and Bloch impedance shown in Fig. 8.17. Theoretical curves of the attenuation and phase constants as well as Bloch impedance are shown for comparison. Since this periodic structure consists of non-symmetrical unit cells, the ABCD-analysis was performed with the following matrices:

$$\begin{bmatrix} A & B \\ C & D \end{bmatrix} = \begin{bmatrix} \cos(\beta_0 \frac{d_c}{2}) & jZ_0 \sin(\beta_0 \frac{d_c}{2}) \\ j\frac{1}{Z_0} \sin(\beta_0 \frac{d_c}{2}) & \cos(\beta_0 \frac{d_c}{2}) \end{bmatrix} \begin{bmatrix} 1 & 0 \\ j\omega C & 1 \end{bmatrix} \begin{bmatrix} \cos(\beta_0 \frac{d_c}{2}) & jZ_0 \sin(\beta_0 \frac{d_c}{2}) \\ j\frac{1}{Z_0} \sin(\beta_0 \frac{d_c}{2}) & \cos(\beta_0 \frac{d_c}{2}) \end{bmatrix} \times $$
$$\begin{bmatrix} \cos(\beta_0 \frac{d_l}{2}) & jZ_0 sin(\beta_0 \frac{d_l}{2}) \\ j\frac{1}{Z_0} sin(\beta_0 \frac{d_l}{2}) & cos(\beta_0 \frac{d_l}{2}) \end{bmatrix} \begin{bmatrix} 1 & j\omega L \\ 0 & 1 \end{bmatrix} \begin{bmatrix} cos(\beta_0 \frac{d_l}{2}) & jZ_0 sin(\beta_0 \frac{d_l}{2}) \\ j\frac{1}{Z_0} sin(\beta_0 \frac{d_l}{2}) & cos(\beta_0 \frac{d_l}{2}) \end{bmatrix}$$
(8.16)

Using the definition of the capacitive and inductive loading factors given in Sections 8.1.1 and 8.2.1 and geometrical dimensions of the unit cell, capacitive and inductive loading factors were calculated to be $LF_C$=1.28 and $LF_L$=0.63 respectively. There is a reasonable agrement between measured and calculated curves of $\beta d$ and $Z_b$. However, since the model assumes a lossless periodic structure, it can not accurately predict the attenuation constant.

Still, the theoretical curve of $\alpha d$ indicates a stopband around 80 GHz, which is in agreement with the S-parameter measurements.

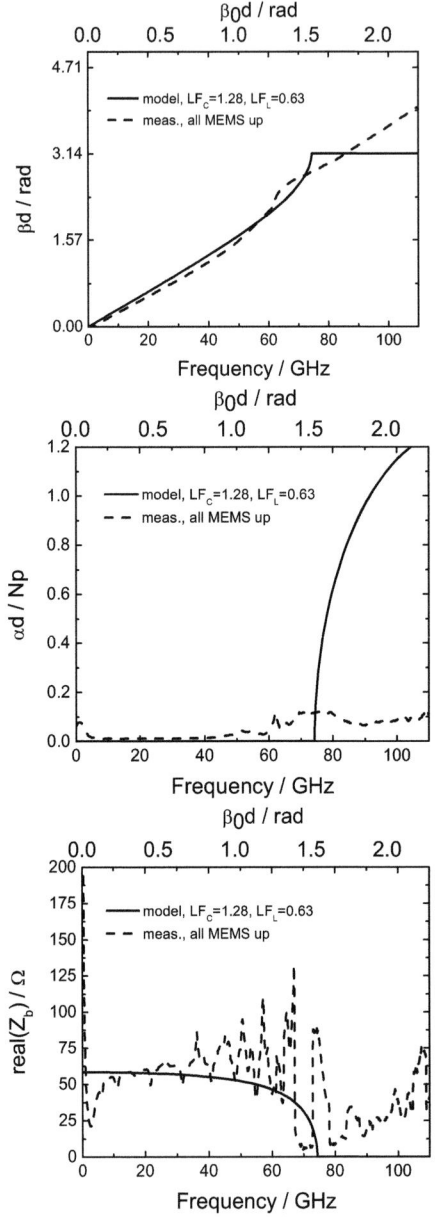

Figure 8.17: Measured and modelled phase delay, attenuation and Bloch impedance of the periodic structure shown in Fig. 8.15.

# Part III

# Circuits Utilizing RFMEMS Periodic Structures

# Chapter 9

# Distributed RFMEMS Phase Shifters

Distributed MEMS transmission line (DMTL) topology is often used to realize RFMEMS phase shifters (see e. g. [1]). A DMTL circuit typically consists of a transmission line, periodically loaded with RFMEMS switchable/tunable capacitors or RFMEMS switchable/tunable inductors. Through MEMS biasing, the phase velocity of the transmission line can be adjusted and thus, different phase delays can be obtained and utilized for the phase shifter operation. General periodic structures used in DMTL applications were described in Part II.

Design of DMTL phase shifters involves careful optimization of the unit cell parameters, as was done using the equations derived in Part II and CAD tools. As shown in Section 8.1.5, low-loss performance is maintained up to about 70 % of the cut-off frequency of the first passband. Above this frequency, the attenuation constant significantly increases and circuit operation at this frequency range should be avoided. Thus, to extend the operation frequency of phase shifters, the Bragg cut-off frequency should be increased. From equations derived in Part II, it follows that this is achieved by shrinking the lateral dimension of the unit cell $d$ and by reducing the loading factor $LF$. This, however, results in the undesirable reduction of the phase delay. In addition, the inherent mismatch also limits allowable MEMS capacitance values.

In the following, several DMTL phase shifters with shunt capacitive and also with a combination of shunt capacitive and series inductive tuning are presented.

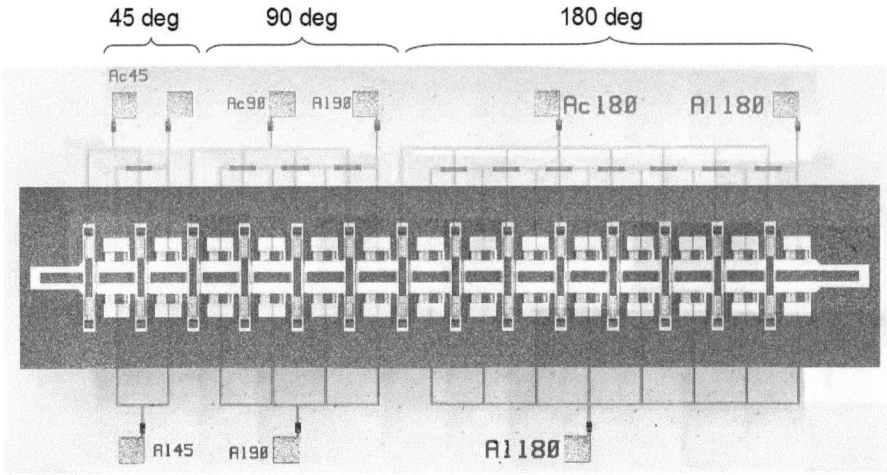

Figure 9.1: Micrograph of the DTML phase shifter with capacitive and inductive tuning. The footprint of the circuit is $6.8 \times 1.4$ mm$^2$ without bias pads.

## 9.1 22 GHz DMTL Phase Shifter with Capacitive and Inductive Tuning

A distinguishing feature of the presented circuit is that the phase delay is adjusted by both shunt RFMEMS capacitors and series RFMEMS inductors. The motivation behind this approach is to overcome the inherent mismatch problem of DMTL phase shifters with only shunt capacitive tuning. The drawback is, however, an increase of the unit cell length, since it is formed by two RFMEMS elements. As a result, the passband cut-off frequency decreases, thus limiting the operation bandwidth of the circuit. Considering typical geometries of RFMEMS components, this approach is suitable at most for microwave frequencies. To utilize it also in millimeter-wave range, more compact MEMS like the ones presented in Section 4.7 must be used.

A micrograph of the described DMTL phase shifter is shown in Fig. 9.1. The unit cell consists of an inductor realized as a MEMS-switchable defected ground structure (DGS) (see Section 5.3) and a MEMS switchable capacitor (see Section 5.4). An SEM image of the unit

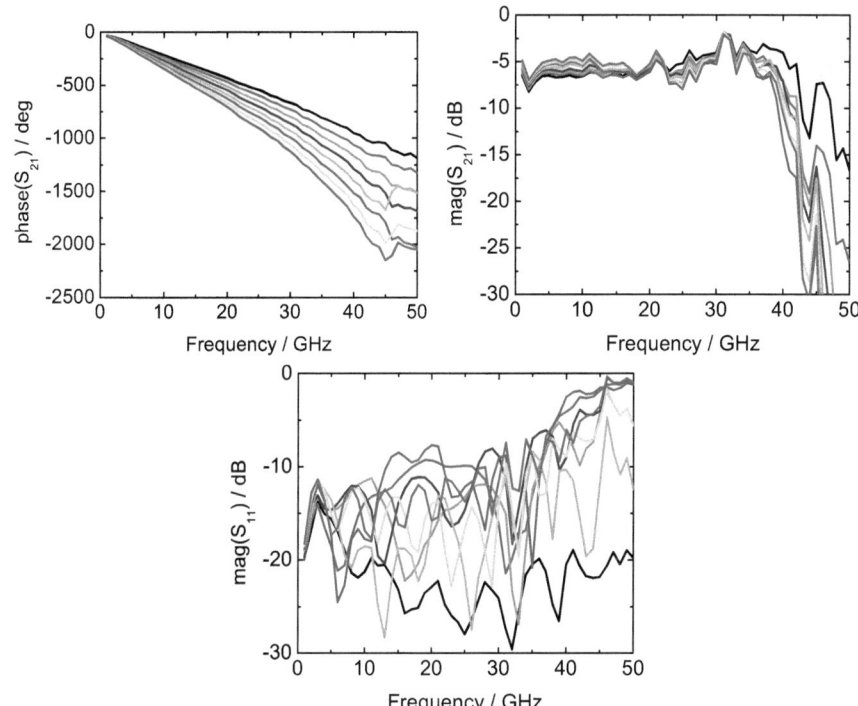

Figure 9.2: Performance of the FBK phase shifter estimated by cascading measured S-parameters of the unit cell in Agilent ADS.

cell was shown in Fig. 5.8. The low phase delay state is obtained when the DGS-cantilever is in the down-state and the capacitive bridge is the up-state. If the bias configuration is reversed, a higher phase delay is produced. This phase shifter is a three-bit 360 deg design. The least significant bit is formed by two unit cells. The two higher bits are composed of four- and eight unit cells respectively. The footprint of the circuit is $6.8 \times 1.4$ mm$^2$ without bias pads.

Unfortunately, measurement of all states of the phase shifters is not possible due to the charging of the thin dielectric layer in shunt capacitive bridges, as described in Section 5.4. Thus, the phase shifter performance was estimated in Agilent ADS by cascading blocks of measured S-parameters of single RFMEMS components. The results are shown in Fig. 9.2. These results predict 360 deg three-bit phase shifter operation at 22 GHz. The phase

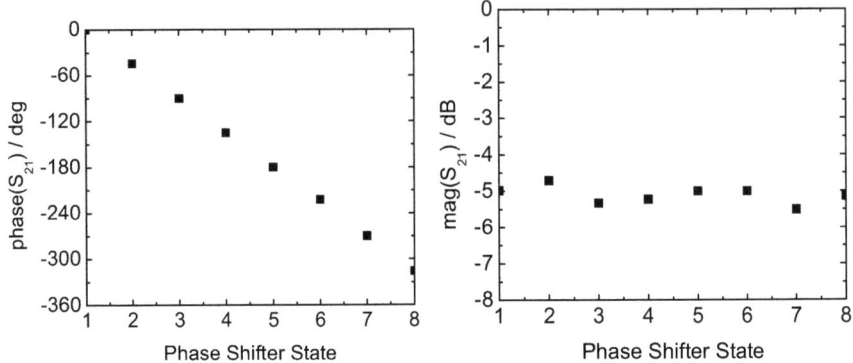

Figure 9.3: FBK phase shifter performance at 22 GHz versus phase shifter state.

increment is about 22.5 deg. The insertion loss of the phase shifter at 22 GHz is better than 5.5 dB and is mainly determined by the losses in the transmission lines. Since in the down-state loading factors differ from each other, the Bloch impedance is not the same as that in the up-state, leading to mismatch and reflections. Still, the return loss at 22 GHz is better than 10 dB.

## 9.2  26 GHz One-port Reflection-Type Phase Shifter with Shunt Capacitive Switches

This phase shifter was designed within the research initiative RARPA (Reflectarrays and Reconfigurable Printed Antennas) of the European Network of Excellence AMICOM, supported by the European Commission. The intended application of the phase shifter is beam-steering of Ka-band reflectarray antennas. In the following the beam-steering principle and specifications of the reflectarray are presented, followed by details of the phase shifter design and experimental results.

## 9.2.1  26 GHz Beam-steerable Reflectarray

Beam-steerable reflectarrays can potentially be used for many applications like low orbit communication satellites, automotive radar systems and remote sensing. Among their advantages are low cost, low mass and improved radiation efficiency as compared to phased array antennas or parabolic reflectors. The higher radiation efficiency of reflectarrays is mainly due to direct feeding from radiating apertures and not through a separate feeding network, which are a significant source of losses in other types of antennas. Control of the radiation direction is achieved by placing a phase shifter next to each radiating element and adjusting the phase of the individual reflected waves accordingly. Here we wish to explore a MEMS-based phase-shifting network due to its low loss and high linearity. The former property is especially important to reduce the receiver noise figure, while the latter is needed to handle high output power in the transmit mode.

Generally, beam-steerable reflectarrays can be divided into two groups: (i) with tunable loads integrated into the radiating elements themselves [63–65] or (ii) with the phase shifters physically separated from the radiating elements [66, 67]. The reflectarray considered here was designed by the University of Perugia within the AMICOM cooperation [67] and it uses the latter approach. It is a microstrip patch reflectarray, slot-coupled to a tunable coplanar load. Its unit cell is schematically shown in Fig. 9.4. As illustrated, the microstrip patches are located on one side of the substrate, while the MEMS-tunable loads are placed on the opposite side. The whole system is fabricated by the Fraunhofer Institute for Silicon Technology (ISiT) on a single high-resistivity silicon wafer processed from both sides. The incident wave will be reflected by the phase shifter with a particular phase delay depending on the phase shifter state. The desired radiation pattern is formed by providing a proper phase distribution at the individual patches across the reflectarray. Thus, a phase shifter with a high resolution and low loss is required. Besides, the phase shifter should be small enough to fit onto one unit cell of the array, whose dimensions are limited by half a wavelength in free space at the operation frequency of 26 GHz (5.77 mm).

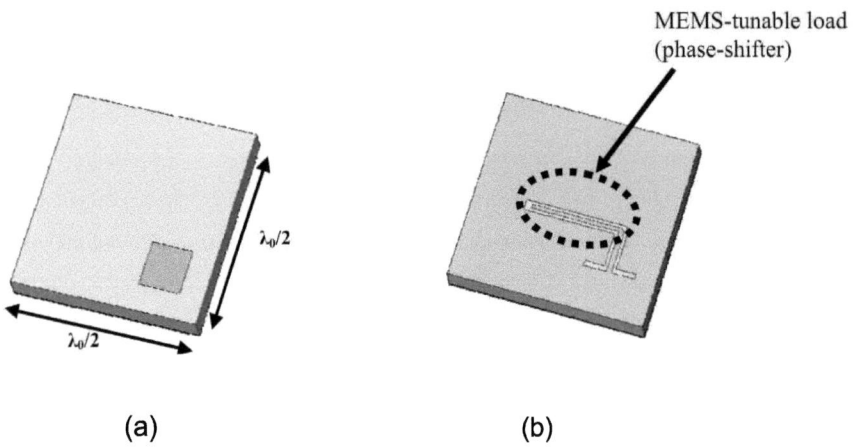

Figure 9.4: Schematic representation of a reflectarray unit cell: (a) microstrip patch on the top side and (b) MEMS-switchable load in coplanar environment on the bottom side

### 9.2.2 Reflection-Type RFMEMS Phase Shifter

Generally, MEMS phase shifters can be realized as distributed MEMS transmission lines (DMTL), switched delay lines or reflection-type topologies. Owing to the tight space constrains, only the reflection-type phase shifter is feasible and is considered in the following. Since the phase shifter is intended to be used in a reflectarray, it is a one-port device sharing the RF input and output. This eliminates any 90 deg hybrids required in case of two-port circuits, making the phase shifter very compact. Besides, due to the reflective nature of the circuit, the physical length of the phase shifter should correspond only to one half of the required electrical length.

An SEM image of the phase shifter is shown in Fig. 9.5. It consists of a coplanar line, loaded with six shunt capacitive MEMS switches, described in Section 4.3. The circuit covers 360 degree in seven steps. With an area of about $1 \times 2$ mm$^2$, the circuit is quite compact as compared to other 360-deg MEMS phase shifters. The unit cell of the phase shifter is formed by a piece of coplanar line of length $d$ and one MEMS switch. Thus, its general description is the same as that of the periodic structure presented in Section 8.1.

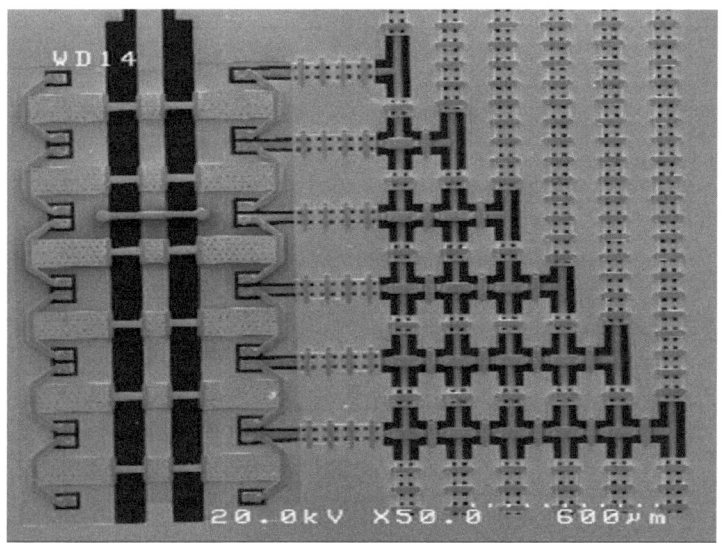

Figure 9.5: SEM image of the reflection-type phase shifter

One end of the coplanar line is used as the RF input/output and the other end is short-circuited. Actuating either of the switches creates a low-impedance path to ground for the RF-signal. The input signal travels till the location of the actuated switch, is reflected there and then travels back to the output. When none of the switches are actuated, the signal is reflected at the short-circuited end of the coplanar line. Ideally, the switch in the down-state would act as a short-circuit at 26 GHz to fully reflect the incoming signal, i.e. the isolation of the switch should be maximized. In practice, an ideal short-circuit can not be obtained. Simulations showed that a low impedance resulting in 15 dB isolation of the switch in the down-state is sufficient for the phase shifter operation.

The main parameter of the phase shifter is its resolution set by the minimum phase shift increment $\phi$. For a frequency $\omega$ not too close to the first passband cut-off, $\phi$ can be approximated by:

$$\phi = 2\beta d = 2\frac{\omega}{V_{ph}}d = 2\omega\sqrt{L'\left(C' + \frac{C_{up}}{d}\right)}$$

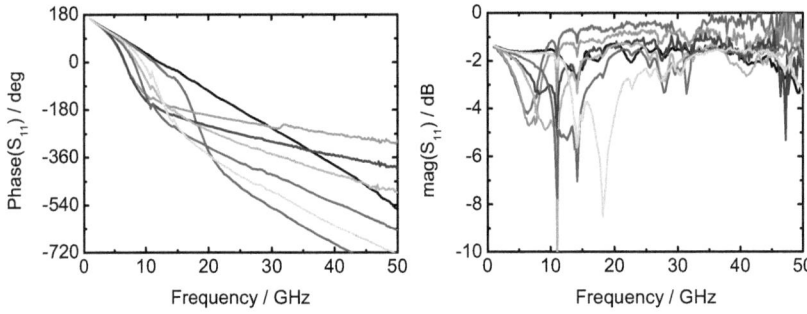

Figure 9.6: Measured $S_{11}$ versus frequency for all states of the reflection-type phase shifter

where $\beta$ is the phase constant, $d$ is the spacing between neighbouring switches, $V_{ph}$ is the phase velocity, $L'$ and $C'$ are the distributed inductance and capacitance of the unloaded coplanar line respectively and $C_{up}$ is the MEMS's up-state capacitance. The factor of two takes into account that the signal will travel back and forth, thus twice the distance between the neighbouring switches $d$. For maximum resolution the spacing $d$ and the up-state capacitance $C_{up}$ should be minimized.

Thus, the spacing $d$ is set to the minimum value of 262 μm given by the technological constraints and the up-state capacitance is chosen as small as possible, yet so that the down-state capacitance is not reduced too much to ensure that the switch provides at least 15 dB isolation, as explained in Section 4.3. With $d = 262$ μm and $C_{up} = 30$ fF, one phase shift segment results in a phase shift of about 51 degrees at 26 GHz. Thus, 360 degrees are covered in seven states provided by six MEMS switches and the short-circuited end of the coplanar line.

Fig. 9.6 illustrates the phase shifter performance up to 50 GHz for all states. At 26 GHz the loss for all states is below 2.5 dB and the overall phase shift is 360 deg. The increased insertion loss at frequencies below 20 GHz can be explained by the fact that the down-state capacitance is not high enough to reflect the signal effectively in this frequency range. Thus, the signal passes through the switch and travels along the short-circuited coplanar line with a significantly reduced phase velocity due to the high loading of the down-state capacitance, as is also clear from Fig. 9.6. Thus, the structure is operated close to the Bragg reflection frequency, leading to high attenuation of the signal at low frequencies. On the other hand,

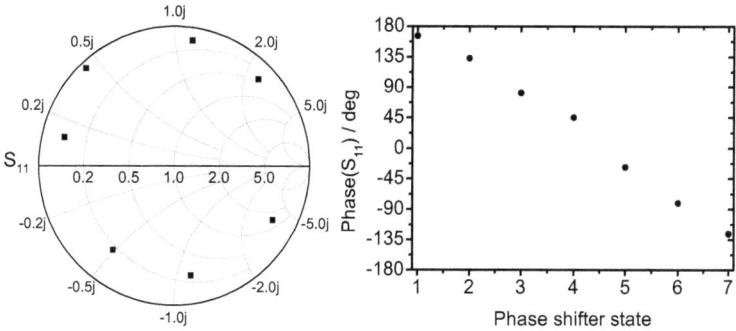

Figure 9.7: Measured $S_{11}$ at 26 GHz for all states of the reflection-type phase shifter

as frequency increases the MEMS bridge approaches an ideal short circuit in the down-state and reflects the signal more effectively. This explains why the insertion loss does not vary significantly between 20 GHz and 45 GHz.

Fig. 9.7 illustrates the measured performance of the phase shifter at 26 GHz. The Smith chart clearly shows that 360 deg are uniformly covered. However, maximum phase error is about 18 deg. Two reasons can be responsible for phase errors. First, there can be slight variations in the up-state capacitances of individual MEMS switches owing to stresses developed in membranes during fabrication. This may cause some membrane deformations and lead to small differences of the up-state capacitances of the neighbouring switches. Besides, phase errors can be introduced, if the down-state capacitances of individual switches are not exactly the same and thus reflect the signal with a slightly different phase. Variations in the down-state capacitances are again caused by slight membrane deformation in the fabrication.

## 9.3 64 GHz DMTL Phase Shifter with Capacitive Tuning

The DMTL phase shifter is based on the periodic structure described in Section 8.1.5 and its SEM-image can be seen in Fig. 8.4. The circuit consists of a coplanar line loaded with eight shunt capacitive MEMS bridges, operated as switched capacitors, which form a distributed MEMS transmission line (DMTL). The MEMS capacitors are presented in Section 4.4. The circuit covers an overall 180 deg phase shift in 8 steps. In the minimum phase shift, i.e.

the zero-degree reference state, all MEMS bridges are in the unbiased position. To obtain higher phase delays, the MEMS capacitors are pulled down one by one, resulting all together in eight discrete phase shifter states in steps of 22.5 deg. The unit cell phase difference of 22.5 deg is obtained at 63.5 GHz, as it can be calculated from the phase delay of eight unit cells in the up- and down-states illustrated in Fig. 8.5.

In most cases, DMTL phase shifters are controlled in a binary fashion, i. e. by grouping the MEMS bridges into bit segments: the least significant bit (LSB), a segment with twice the LSB phase delay, a segment with four times the LSB phase delay and so on depending on the number of bits $n$. Thus, $n$ controlled signals are used to set the state of the individual bits. The disadvantage of this approach is that it results in very inhomogeneous electromagnetic properties along the structure due to different Bloch impedances of the unit cell in the up- and down-states of the MEMS capacitor, leading to increased reflection losses. To overcome this problem, the thermometer, i. e. base one, code was used in [21] instead of the binary one, that resulted in a significantly improved performance. Thus, the thermometric mode of the phase shifter control was used here as well. This approach is only suitable for phase shifters with a small number of states, otherwise the control circuit complexity may be high.

The measured S-parameters of the phase shifter versus frequency are shown in Fig. 9.8. In states with higher phase shift, one can see a sharp increase of the insertion and return loss above 91 GHz, which is the cut-off frequency of the first passband as it was calculated in Section 8.1.5. Another important observation is that the phase delay becomes highly nonlinear, as the frequency is approaching the passband cut-off and beyond it. Fig. 9.9 illustrates the performance of the phase shifter at the indented operation frequency of 63.5 GHz. The phase delay increases in discrete steps of about 22.5 deg and the maximum phase shift is 180 deg. The insertion loss at 63.5 GHz is below 4.7 dB and the return loss is better than 8 dB for all phase shifter states.

The phase shifter can be modelled as a periodic structure with shunt capacitive loading according to the theory described in Section 8.1. As calculated in Section 8.1.5, the measured up- and down-state loading factors $LF_C$ are 0.54 and 1.18 respectively. Fig. 8.6 illustrates, that 63.5 GHz is slightly below the transition frequency, where the down-state attenuation increases considerably. Also, the down-state Bloch impedance starts to decrease, indicat-

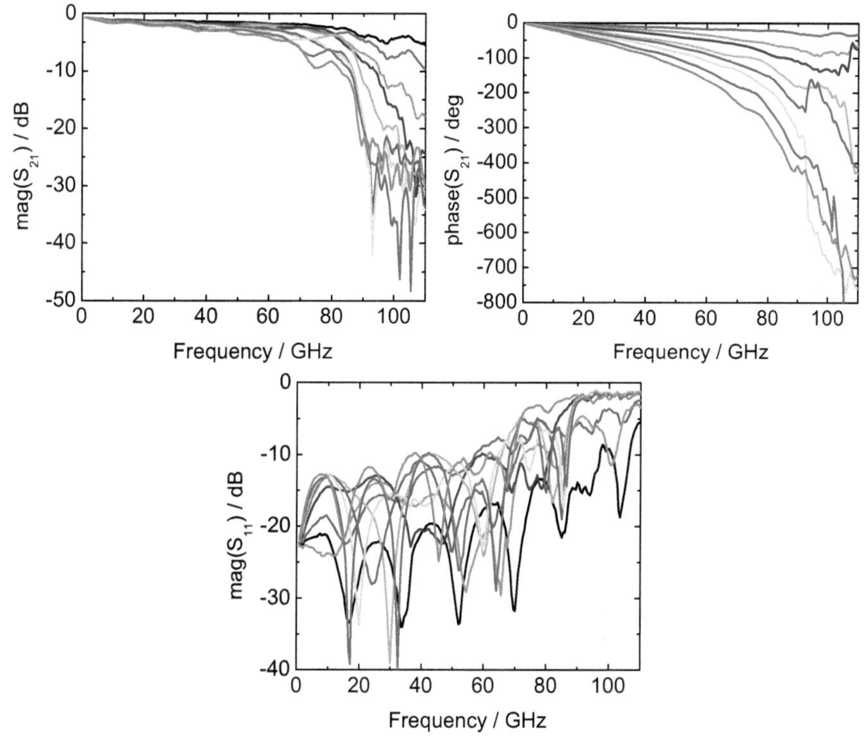

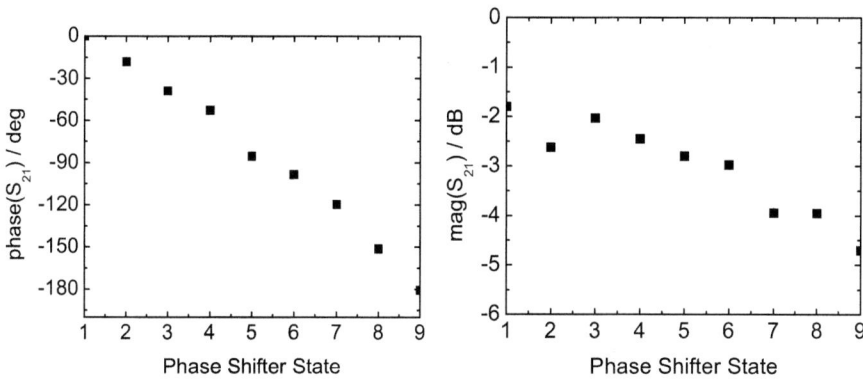

Figure 9.8: Measured S-parameters versus frequency for all states of the phase shifter

Figure 9.9: Measured magnitude and phase of $S_{21}$ at 64 GHz for all states of the phase shifter

ing larger reflection losses. Thus, 63.5 GHz approaches the maximum practical operation frequency of the phase shifter. This corresponds to the DMTL phase shifter operation at about 70 % of the first passband cut-off frequency (91 GHz), meaning a considerably increased bandwidth as compared to the generally recommended operation at 30 % of the cut-off frequency [6].

## 9.4 82 GHz DMTL Phase Shifter with Capacitive Tuning

The operation principle of this phase shifter is identical to the V-band phase shifter discussed in Section 9.3, but its operation frequency is shifted to the E-band due to different properties of the underlying periodic structure. The SEM-image of the E-band phase shifter can be seen in Fig. 8.7. The circuit is designed as a coplanar transmission line loaded with eight RFMEMS switchable capacitors presented in Section 4.5. The unit cell length of the E-band phase shifter is about 20 % smaller than that of the V-band design (300 μm vs. 360 μm). Considering that both circuits have similar loading factors, the passband cut-off frequency is about 20 % higher (109 GHz vs. 91 GHz) for the smaller unit cell design. This explains why the phase shifter with a shorter unit cell can operate at higher frequencies as compared to the V-band design.

As in case of the phase shifter presented in Section 9.3, thermometer coding is used to reduce the inter-bit reflections. The lowest phase delay state corresponds to the case when none of the MEMS capacitors are actuated. The largest phase delay is obtained when all MEMS are in the down-state. The measured S-parameters of the phase shifter versus frequency are shown in Fig. 9.10. At 82 GHz the circuit acts as a 180 deg phase shifter covering the phase in eight steps of 22.5 deg each. The phase delay and insertion loss of the phase shifter versus the phase shifter state at 82 GHz are shown in Fig.9.11. The insertion loss at 82 GHz is below 4.7 dB and the return loss is better than 7 dB for all phase shifter states.

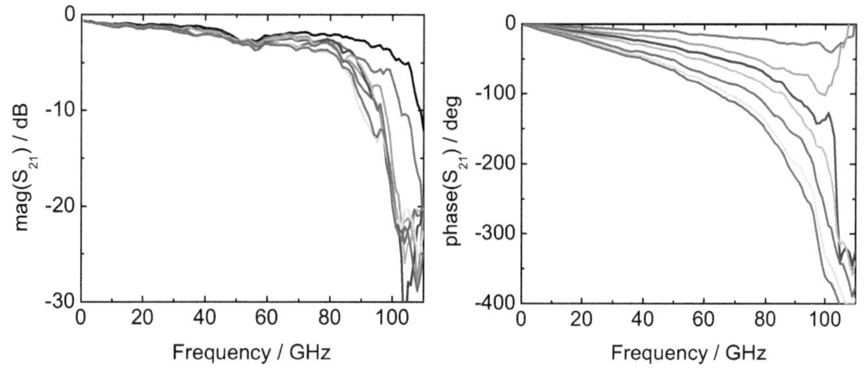

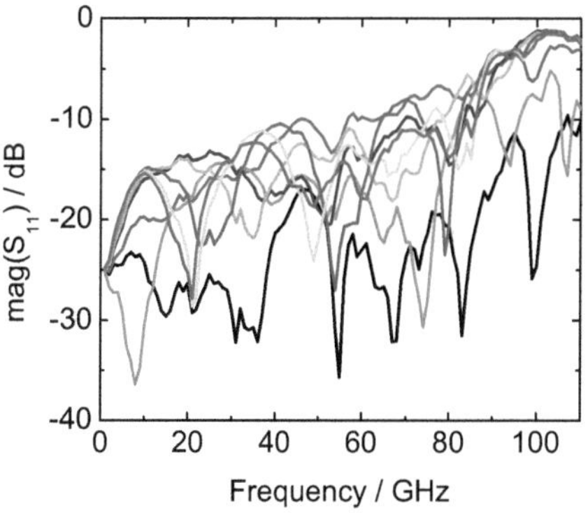

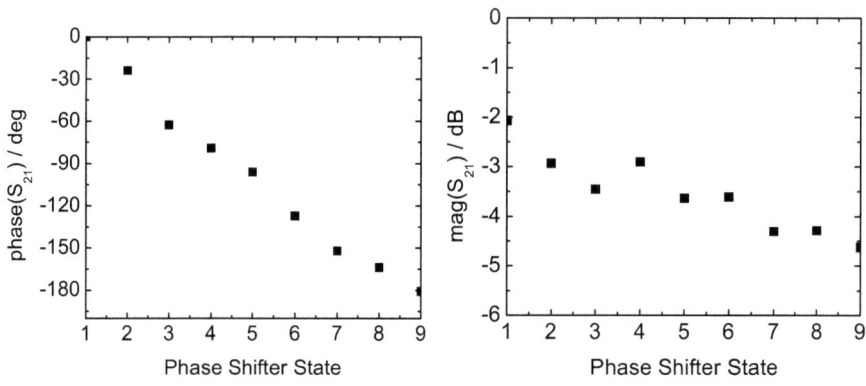

Figure 9.10: Measured $S_{11}$ versus frequency for all states of the phase shifter

Figure 9.11: Measured magnitude and phase of $S_{21}$ at 82 GHz for all states of the phase shifter

# Chapter 10

# Distributed RFMEMS Impedance Tuners

Distributed RFMEMS impedance tuners are designed as a transmission line typically loaded with switchable MEMS capacitors (e. g. [3, 68]), although switchable inductors or their combination with switchable capacitors can be used as well. Distributed impedance tuners take advantage of the impedance variation resulting from actuation of RFMEMS components. This clearly differentiates them from distributed RFMEMS phase shifters, which rely on the change in the phase velocity. It has been shown [69] that distributed topologies result in better power handling as compared to stub-based impedance tuners. In the following, several distributed impedance tuners with capacitive and inductive RFMEMS switching are presented.

## 10.1 Ka-Band Impedance Tuner with Capacitive and Inductive Tuning

The distinguishing feature of this impedance tuner is that it utilises both capacitive and inductive RFMEMS tuning, resulting in large VSWR and uniform impedance coverage of the Smith Chart. The circuits was designed using FBK RFMEMS switchable inductors presented in Sections 5.3 and switchable shunt capacitors described in Section 5.4. The underlying periodic structure is the same as considered in Section 8.3.5 and also used for the phase shifter design in Section 9.1.

The micrograph of the fabricated impedance tuner is shown in Fig. 10.1. The unit cell consists of one MEMS capacitor, one MEMS inductor and interconnecting transmission lines. The circuit is composed of seven cascaded unit cells resulting in $2^{14}$=16,384 different impedance states depending on the bias points of the individual MEMS components. In principle, a good impedance coverage can be obtained also with smaller number of "bits", where each bit represents one MEMS component. So, eight bits (four unit cells) or ten bits (five unit cells) may be sufficient depending on the impedance coverage requirements and on the tolerable complexity of the bias network.

As explained in Section 5.4, the shunt capacitors are susceptible to dielectric charging and thus, the measurement of the structures containing a large number of these components, e. g. this impedance tuner, was not feasible given the state of development of the underlying RFMEMS process. Thus, as for the phase shifter, the performance of the impedance tuner has been simulated in Agilent ADS by cascading the measured S-parameter blocks of individual MEMS components. The results are shown in Fig. 10.2. The impedance coverage is simulated for two cases: eight bits (four unit cells) or ten bits (five unit cells). It can be seen, that five unit cells are sufficient to obtain an dense and uniform impedance coverage. The best performance is obtained for 35-45 GHz.

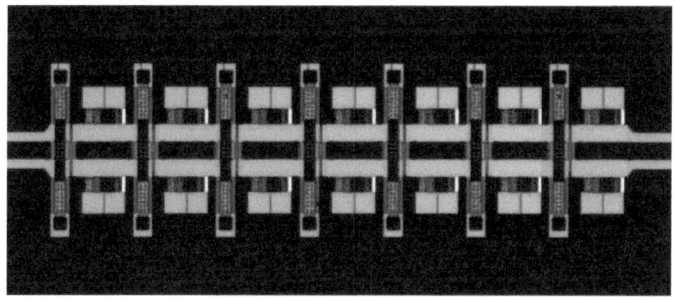

Figure 10.1: Micrograph of the FBK DTML impedance tuner with capacitive and inductive tuning. The circuit dimensions are 3912×1398 µm$^2$.

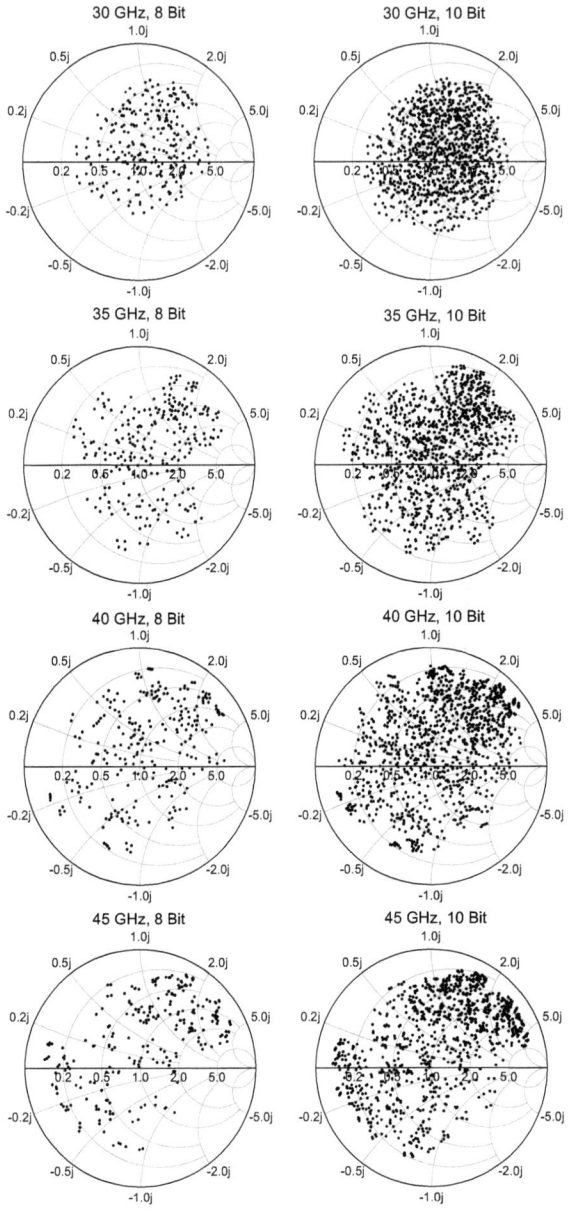

Figure 10.2: Impedance coverage of DMTL impedance tuner with capacitive and inductive tuning using RFMEMS capacitors and inductors presented in Sections 5.3 and 5.4.

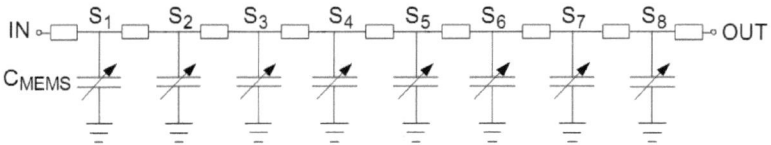

Figure 10.3: Equivalent circuit of the impedance tuner with capacitive tuning

## 10.2 W-band Impedance Tuner with Capacitive Tuning: Design 1

This impedance tuner is based on the periodic structure shown in Fig. 8.7. It is in principle the same circuit, as used for the 82 GHz phase shifter design described in Section 9.4. As mentioned above, the circuit consists of a coplanar transmission line, periodically loaded with eight RFMEMS switchable capacitors described in Section 4.5 with the corresponding equivalent circuit shown in Fig. 10.3. Each MEMS capacitor is actuated independently, thus changing the input impedance of the circuit. Thus, with eight two-state capacitors, the circuit can synthesize $2^8 = 256$ different impedances.

The impedance tuner performance has been measured at VTT Millilab, Finland, using an automated measurement setup. The resulting reflection coefficients for all impedance states are shown in Fig. 10.4. At frequencies below 50 GHz the impedance of the shunt MEMS capacitors is relatively large and their effect is not so pronounced. Thus, the generated impedances are close to the center of the Smith Chart. As frequency increases, much higher VSWR can be generated. The best impedance coverage, i.e. uniform and with large VSWR, is achieved at 80-100 GHz. Finally, at very high frequencies (above 100 GHz), the impedance coverage becomes not uniform again, since the MEMS impedance is becoming too small and only low impedances can be generated.

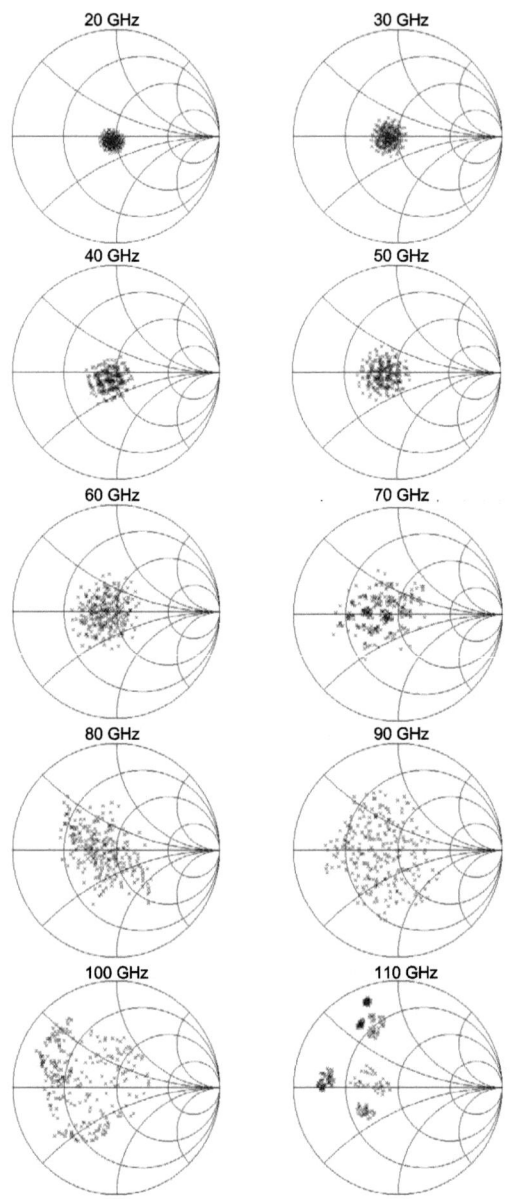

Figure 10.4: Measured impedance coverage of DMTL impedance tuner using RFMEMS capacitors presented in Section 4.5

## 10.3 W-band Impedance Tuner with Capacitive Tuning: Design 2

To achieve a better impedance coverage of the Smith Chart, a second impedance tuner was designed as an extended version of the circuit presented above. It is the same periodic structure used for the 63 GHz phase shifter design in Section 9.3. Similar to the previous impedance tuner, the circuit is based on a switchable capacitive loading of a transmission line, but utilizes tri-state MEMS capacitors presented in Section 4.4. Hence, its equivalent circuit is identical to the one shown in Fig. 10.3.

An SEM image of the circuit is shown in Fig. 8.4. It consists of eight tri-state RFMEMS capacitors loading a coplanar transmission line. Thus, the circuit can generate $3^8 = 6561$ different impedance states, resulting in a more dense Smith Chart coverage. In addition, this RFMEMS capacitor is more robust against possible difference in membrane stresses and thus results in higher yield, particularly important for circuits, which employ a large number of such components.

Measured reflection coefficient of the impedance tuner is shown in Fig. 10.5. As with the previous tuner, these measurements were performed at VTT Millilab using an automated measurement system. Due to large number of possible impedance states, the MEMS capacitors were switched only between the low and intermediate capacitance states, resulting in $2^8 = 256$ measured states of the tuner. As Fig. 10.5 shows, the optimum frequency band is 80-100 GHz, when the impedance tuner generates the largest VSWR and the produced impedance covers the Smith Chart uniformly.

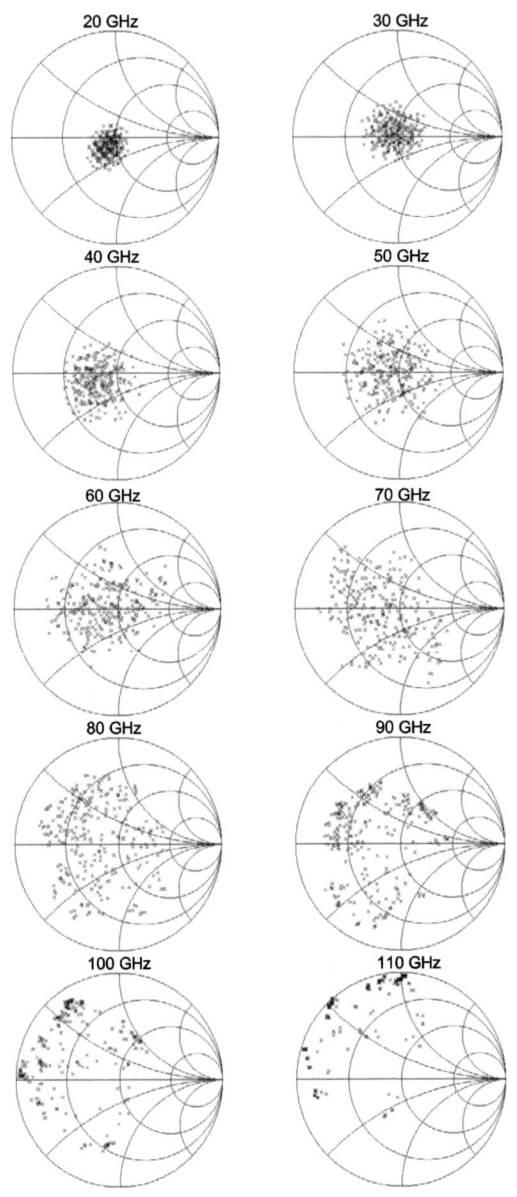

Figure 10.5: Measured impedance coverage of DMTL impedance tuner using RFMEMS capacitors presented in Section 4.4

# Conclusions and Outlook

# Chapter 11

# Conclusions

Main outcomes of this thesis are summarized below.

### Part 1

- ISiT and FBK RFMEMS portfolios were extended with several RFMEMS switches, shunt- and series switchable capacitors as well as a switchable inductor based on defected ground structures.

- Switching and release dynamics were analysed experimentally and release oscillations were approximated with the damped oscillation model. Analytical results for the switching time equation derived in [39] were validated with the MEMS components developed in this work.

- It was confirmed experimentally that the release time is not set uniquely by the release oscillations, but is also dependent on the frequency of the RF signal.

### Part 2

- Periodic structures with shunt capacitive loading, series inductive loading and their combination were described using ABCD-analysis.

- It was shown that while calculating phase constant of a periodic structure, care should be taken to select correct values of the phase delay $\beta d$. Simply taking the principal value of the inverse cosine function leads to incorrect (unphysical) results, sometimes reported in the literature.

- Using the loading factor concept from [5], it was shown that the A-coefficient of the ABCD-matrix for periodic structures with shunt capacitive is identical to that of series inductive loading resulting in equal propagation constants.

- It was shown analytically that by using both shunt capacitive and series inductive loading with equal loading factors, the issue of inherent mismatch of DMTL phase shifters can be eliminated for a broad frequency range.

- Accurate approximate equations for cut-off frequencies for all pass- and stopbands of periodic structures with solely shunt capacitive or series inductive loading were derived.

- The derived equations for propagation constant and Bloch impedance of periodic structures with shunt capacitive as well as with both shunt capacitive and series inductive loading were verified experimentally up to 110 GHz.

**Part 3**

- Ka-, V- and W-band phase shifters with capacitive and capacitive-inductive tuning were designed based on the developed RFMEMS periodic structures.

- Operation frequency of phase shifters was extended from the recommended 30 % [6] to 70 % of the cut-off of the first passband.

- Ka- and W-band impedance tuners also with capacitive as well as capacitive-inductive tuning were developed.

- Good impedance coverage at V- and W-bands was obtained, making the circuits suitable for load-pull and noise parameter measurement systems.

# Chapter 12

# Outlook

## 12.1 Millimeter-wave RFMEMS and Their Monolithic MMIC Integration

RFMEMS components are typically fabricated in non-standardized dedicated processes, not compatible with MMIC. Thus, only a hybrid integration with active circuits is possible. However, due to low loss and inherently high linearity even at elevated frequencies, RFMEMS are particularly important for higher millimeter-wave applications., where hybrid integration is not feasible due to associated parasitic and increased size.

Thus, to extend millimeter-wave MMIC's portfolio with RFMEMS, a fully monolithic fabrication must be developed. It is an approach pursued by very few groups and IC foundries worldwide.The French foundry OMMIC is working towards integration of MEMS into a high-frequency compound semiconductor MMIC technology. Here the focus is on developing a series RFMEMS switch with a metal-to-metal contact, operating best at microwave frequencies. Suggestions for relevant RFMEMS circuits like tunable matching networks, phase shifting elements and switching networks have been presented recently [70–72]. In Italy, Selex Systemi Integrati has developed MEMS components using its GaN MMIC technology [73]. The targeted frequency range is again below 50 GHz. IHP Microlectronics embedded an RFMEMS switch into the backend-of-line (BEOL) of its advanced Si/SiGe BiCMOS process [74–78]. IHP Microlectronics aims at mm-wave frequency applications and the switch is optimized for W-band.

Still, there is little research on monolithic MEMS integration and only few experimental results are available for upper millimeter-wave frequencies, where RFMEMS can provide the highest benefit to MMICs. Besides, so far no effort is made to investigate actual MMICs with RFMEMS operating at W-band and beyond. To close this gap, monolithically-integrated MEMS components new circuits concepts arising with MMIC-compatible RFMEMS for upper millimeter-waves should be developed.

## 12.2 Monolithic Integration of RFMEMS into the mHEMT Process of Fraunhofer IAF

GaAs technology is particulary suitable for RFMEMS integration due to the existing air-bridge process, as was also suggested in [79]. Recently, a collaboration between Ulm University, Fraunhofer IAF and KIT demonstrated the feasibility of RFMEMS integration into 100 nm mHEMT technology of IAF by fabricating an RFMEMS switchable series capacitor. The MEMS were processed in the frame of multi-project wafer (MPW) runs together with active MMICs, highlighting their technological compatibility. A simplified illustration of the layers in the mHEMT MMIC process at IAF is shown in Fig. 12.1. The MEMS use a specific combination of the following technological modules:

- MET1: 300 nm gold for bottom electrode of the MEMS capacitive switch
- SiN: dielectric normally used for MIM-capacitors
- METG: 2.7 µm gold for low-loss MEMS membranes
- Wet etching of photoresist to release MEMS within the standard air-bridge process
- MESA and thin-film NiCr resistors for high-resistivity layers for DC-bias lines

The SEM-image of a first prototype and its schematic cross-section are shown in Fig. 12.2. The developed RFMEMS series capacitive cantilever is highly compact with the footprint of only $17 \times 120$ µm$^2$, which is much smaller than that of many state-of-the-art MEMS. Measured S-parameters up to 120 GHz are presented in Fig. 12.3. The actuation voltage is 35 V, at which the electrostatic force exceeds the spring force and the membrane snaps

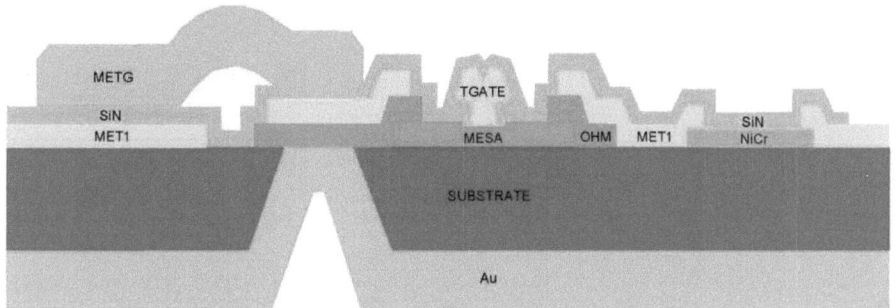

Figure 12.1: Layer stack of the standard mHEMT process at the Fraunhofer IAF. MEMS fabrication uses MET1, SiN, METG, MESA, NiCr and wet etching of the photoresist layer.

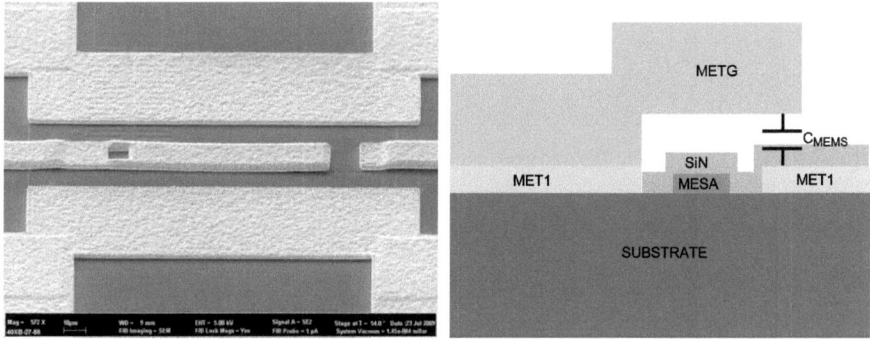

Figure 12.2: SEM-image (left) and a schematic cross-section of the RFMEMS cantilever (right) fabricated in the mHEMT technology of IAF.

down. This leads to a change in capacitance, indicated by a large difference in the magnitude of $S_{21}$. The measured up-state capacitance is about 4.5 fF, which is close to simulation. However, in the down-state, the measured capacitance is only 10 fF, which is much smaller than the simulated 45 fF.

The reason for a low on/off capacitance ratio is the following. It was found out that due to internal stresses in the gold membrane, the cantilever's tip was bended upwards, leading to a small air-gap even in the down-state, which considerably reduced the capacitance. This problem can be solved by redesigning the structure in a way that the electrostatic force is applied also at the tip of the cantilever, ensuring good contact in the down-state. After a

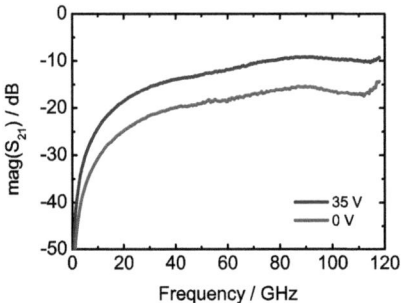

Figure 12.3: Experimentally determined transmission coefficient ($|S_{21}|$) of the first RFMEMS cantilever prototype in mHEMT technology. The actuation voltage is 35 V. The measured on/off capacitance ratio is smaller than expected due to the upwards bending of the cantilever's tip, leading to a small air-gap in the down-state. This problem can be solved in a mechanical redesign.

mechanical redesign, this component will be used as switched series capacitor in a tunable matching network of a dual-band MMIC.

The successful fabrication of the first prototypes of switched capacitor led to a new design of a G-band RFMEMS switch using the mHEMT technology. The purpose was to evaluate whether RFMEMS provide advantages at upper millimeter-wave frequencies, where transistors parasitics considerably increase. With the help of extensive electromagnetic (EM) simulations in Sonnet and mechanical analysis in ANSYS, RFMEMS SPST and SPDT switches for G-band were designed [80]. To obtain a realistic estimation of achievable losses, a very accurate 2.5 D EM simulation was performed and the results are shown in Fig. 12.5. At 195 GHz, the Sonnet simulations indicate 0.9 dB of insertion loss and 34 dB of isolation for the SPST and 1.9 dB of insertion loss and 40 dB of isolation for the SPDT switch. This is a considerable improvement as compared to the mHEMT SPDT recently designed by IAF having an insertion loss of 2.5 dB and isolation of 18 dB at a lower frequency of 140 GHz. The layout of the SPDT consisting of two SPST switches with two $\lambda/4$ transformers is shown in Fig. 12.4. The next step is to fabricate the designed G-band SPST and SPDT switches and test them experimentally.

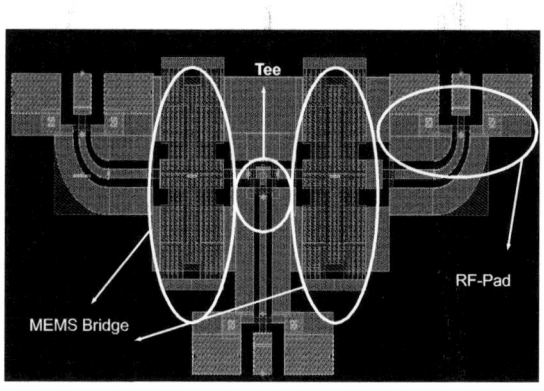

Figure 12.4: Layout of the G-band SPDT RFMEMS switch in mHEMT technology

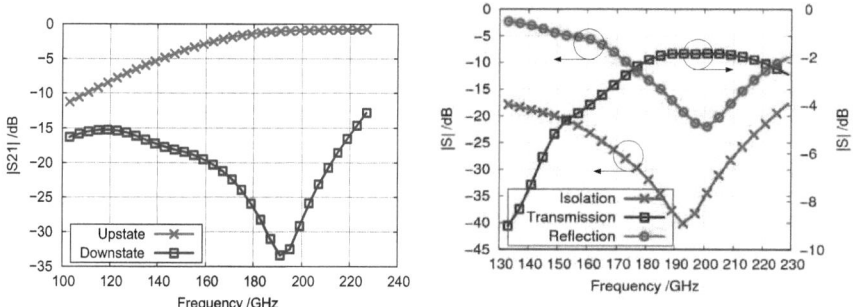

Figure 12.5: Transmission and reflection coefficients ($|S_{21}|$ and $|S_{11}|$) obtained by Sonnet simulations for the SPST switch (left) and SPDT switch (right).

The initial results are very promising, confirming two important points. First of all, it is feasible to fabricate RFMEMS as a part of the already existing mHEMT process. Equally important is the fact that employing RFMEMS at upper millimeter-wave frequencies can indeed significantly improve system performance. However, technological issues like variable bending of membranes due to internal stresses and adhesion forces, and breakdown of GaAs for particularly high DC-voltages. need still to be addressed before sub-mm wave MMICs incorporating RFMEMS can be conclusively demonstrated.

# Appendix

# Appendix A

# Approximate Solutions of Equation $\tan(x) = -\xi x$

Approximate analytical solutions for the equation $\tan(x) = -\xi x$ are derived in the following. The derivation algorithm was originally proposed in [81] for the special case of $\tan(x) = x$ and the algorithm is extended here for the general case for a negative coefficient $-\xi$ (where $\xi$ is positive). In the context of the problem described in Sec 8.1.4, $x$ implies the phase delay of the unloaded line $\frac{\beta_0 d}{2}$ and $\xi$ is the loading factor $LF$. Thus, in most of practical cases, the coefficient $\xi$ is rather small and varies between zero and three.

To find the approximate solutions, first of all the equation $\tan(x) = -\xi x$ is re-written as a system:

$$\begin{cases} y = -\xi x \\ y = \tan(x) \end{cases} \quad \text{(A.1)}$$

The equations A.1 are plotted in Fig. A.1 for three practical cases for $\xi$:

- $\xi = 0.1$ (small MEMS loading)
- $\xi = 1$ (medium MEMS loading)
- $\xi = 3$ (high MEMS loading)

Due to the physical nature of the problem, only positive $x$ are of interest. As Fig. A.1 shows, there is an infinite number of solutions. For higher branches of $\tan(x)$, the abscissas of

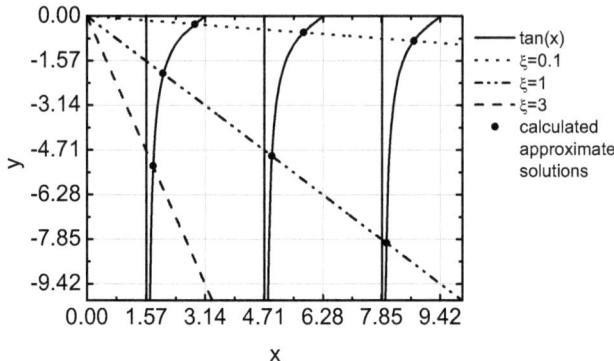

Figure A.1: Graphical representation of the system of equations A.1 and its approximate solutions after one iteration

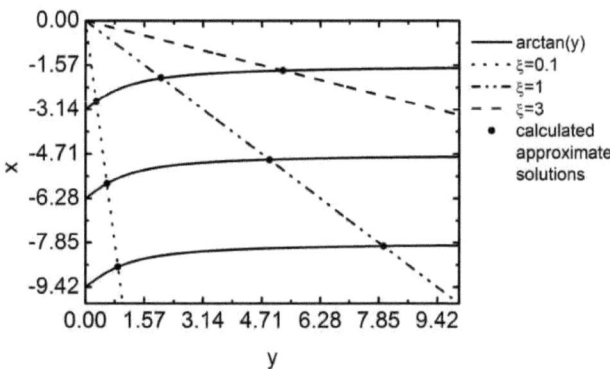

Figure A.2: Graphical representation of the system of equations A.2 and its approximate solutions after one iteration

solutions approach $(2n-1)\frac{\pi}{2}$ with $n$ - non-zero integer, and the corresponding ordinates approach $-\xi(2n-1)\frac{\pi}{2}$. Approximate solutions can be found by replacing the tangent function with its Taylor expansion in the vicinity of these points. However, $\tan(x)$ has break points at $(2n-1)\frac{\pi}{2}$, making it unsuitable for the Taylor expansion. Thus, an inverse of Eq. A.1 can be used instead, since the inverse tangent is a continuous function:

$$\begin{cases} x = -\dfrac{y}{\xi} \\ x = \arctan(y) \end{cases} \quad (A.2)$$

As for the tangent function, equations of the system A.2 are plotted in Fig. A.2 for the same values of $\xi$. Since the inverse tangent is continuous, its principal branch can be approximated in the vicinity of some point $a$ using the Taylor expansion:

$$\arctan(y) = \arctan(a) + \frac{1}{1+a^2}(y-a) + \ldots \quad (A.3)$$

The approximation points are chosen as $a = \xi(2n-1)\frac{\pi}{2}$ to assure that there is a solution $y$ in the vicinity of each $a$, i.e. $|y - a|$ is small. Thus, first two terms of the Taylor expansion are sufficient for a good approximation. It should be noted that the existence of a solution in the vicinity of $a = \xi(2n-1)\frac{\pi}{2}$ is particularly true for higher branches, but also for lower branches it is a sufficient assumption, as shown further down. Besides, to avoid ambiguity, the multi-valued $\arctan$ function is restricted to its principal value. Hence, to obtain the correct corresponding solution for each $a$, a factor of $n\pi$ should be subtracted from the principal value:

$$-\frac{y}{\xi} \approx \arctan(a) - n\pi + \frac{1}{1+a^2}(y-a) \quad (A.4)$$

or equivalently:

$$y \approx -\xi \frac{[\arctan(a) - n\pi][1+a^2] - a}{1 + a^2 + \xi} \quad (A.5)$$

The wanted approximate solutions $x$ are readily found by dividing $y$ obtained from Eq. A.5 by $-\xi$. The calculated values are plotted in Fig. A.1 and Fig. A.2, which indicate a very good accuracy of the solutions for all practical values of $-\xi$.

To increase the accuracy further, this procedure can be repeated once more, but now with application of the Taylor expansion in the vicinity of the approximate solutions $y$ found in the first iteration, i.e. using them in place of $a = \xi(2n-1)\frac{\pi}{2}$. Table A.1 summarizes the results, obtained after first and second iterations for different values of $\xi$. As can be seen, a very high accuracy below $10^{-2}$ % is achieved after the second iteration for the considered values of $\xi$.

Table A.1: Approximate solutions of equation $\tan(x) = -\xi x$

| $\xi = 0.1$ | | | | | | | | |
|---|---|---|---|---|---|---|---|---|
| n | $x_1$ | $-0.1x_1$ | $\tan(x_1)$ | $\delta_1, \%$ | $x_2$ | $-0.1x_2$ | $\tan(x_2)$ | $\delta_2, \%$ |
| 1 | 2.86 | -0.286 | -0.28931 | 1.14 | 2.8628 | -0.28628 | -0.28628 | $<10^{-5}$ |
| 2 | 5.7573 | -0.5757 | -0.58039 | 0.8 | 5.7606 | -0.57606 | -0.57606 | $<10^{-5}$ |
| 3 | 8.7063 | -0.8706 | -0.8744 | 0.43 | 8.7083 | -0.87083 | -0.87083 | $<10^{-5}$ |
| 4 | 11.702 | -1.1702 | -1.17276 | 0.22 | 11.702 | -1.17027 | -1.17027 | $<10^{-6}$ |
| $\xi = 1$ | | | | | | | | |
| n | $x_1$ | $-x_1$ | $\tan(x_1)$ | $\delta_1, \%$ | $x_2$ | $-x_2$ | $\tan(x_2)$ | $\delta_2, \%$ |
| 1 | 2.0108 | -2.0108 | -2.1240 | 5.63 | 2.0287 | -2.0287 | -2.0289 | $<10^{-2}$ |
| 2 | 4.9129 | -4.9129 | -4.9214 | 0.17 | 4.9132 | -4.9132 | -4.9132 | $<10^{-6}$ |
| 3 | 7.9786 | -7.9786 | -7.9806 | 0.02 | 7.9787 | -7.9787 | -7.9787 | $<10^{-8}$ |
| 4 | 11.086 | -11.086 | -11.0863 | $<10^{-2}$ | 11.086 | -11.086 | -11.086 | $<10^{-10}$ |
| $\xi = 3$ | | | | | | | | |
| n | $x_1$ | $-3x_1$ | $\tan(x_1)$ | $\delta_1, \%$ | $x_2$ | $-3x_2$ | $\tan(x_2)$ | $\delta_2, \%$ |
| 1 | 1.75596 | -5.26789 | -5.33865 | 1.33 | 1.75816 | -5.27449 | -5.2745 | $<10^{-3}$ |
| 2 | 4.78197 | -14.3459 | -14.34902 | 0.02 | 4.78198 | -14.34595 | -14.34595 | $<10^{-9}$ |
| 3 | 7.89617 | -23.68851 | -23.68919 | 0.02 | 7.89617 | -23.68851 | -23.68851 | $<10^{-11}$ |
| 4 | 11.0258 | -33.07739 | -33.07764 | $<10^{-3}$ | 11.0258 | -33.07739 | -33.07739 | $<10^{-12}$ |

# Appendix B

# Approximate Solution of Equation $\cot(x) = +\xi x$

Approximate solutions for this equation are found with a similar algorithm used for the tangent function described in Appendix A. As before, $x$ implies the phase delay of the unloaded line $\frac{\beta_0 d}{2}$ and $\xi$ is the loading factor $LF$. First of all, the equation is re-written as a system:

$$\begin{cases} y = \xi x \\ y = \cot(x) \end{cases} \quad (B.1)$$

The equations B.1 are plotted in Fig. B.1 for the same three values for $\xi$ (0.1, 1, 3). Again, only positive $x$ are of interest. As before, the inverse of Eq. B.1 is used for the Taylor expansion, since the inverse cotangent function is continuous:

$$\begin{cases} x = \dfrac{y}{\xi} \\ x = arccot(y) \end{cases} \quad (B.2)$$

The principal branch of the inverse cotangent can be approximated in the vicinity of some point $a$ using the Taylor expansion:

$$arccot(y) = arccot(a) - \frac{1}{1+a^2}(y-a) + ... \quad (B.3)$$

Fig. B.1 shows that there is infinite number of solutions. For higher branches of $\cot(x)$, the abscissas of solutions approach $n\pi$ with $n$ - non-zero integer, and the corresponding ordinates approach $-\xi n\pi$. However, for lower branches the solutions significantly deviate from these points. A good compromise for all branches is achieved by selecting $a = (2n - 1)\frac{\pi}{4}$. However, two iterations of the algorithm are necessary, as will be shown below.

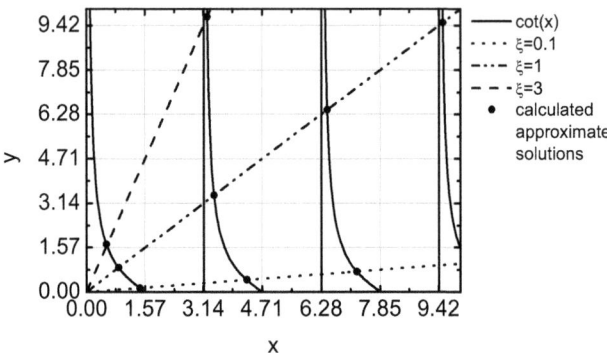

Figure B.1: Graphical representation of the system of equations B.1 and its approximate solutions after two iterations

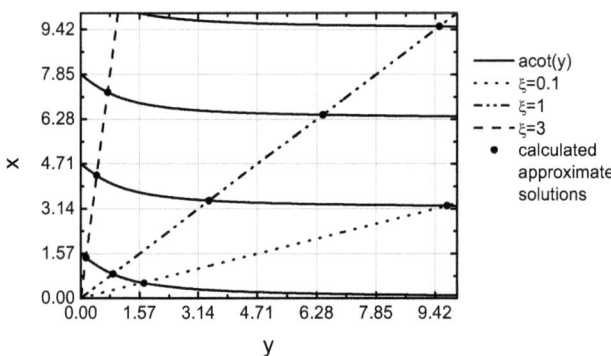

Figure B.2: Graphical representation of the system of equations B.2 and its approximate solutions after two iterations

Similar to the tangent function, the multi-valued $arccot$ function is restricted to its principal value. Thus, a correction factor $(n-1)\pi$ should be added to its principal value:

$$\frac{y}{\xi} \approx arccot(a) + (n-1)\pi - \frac{1}{1+a^2}(y-a) \tag{B.4}$$

or equivalently:

$$y \approx \xi \frac{[arccot(a) + (n-1)\pi](1+a^2) + a}{1 + a^2 + \xi} \quad \text{(B.5)}$$

The wanted approximate solutions $x$ are readily found by dividing $y$ obtained from Eq. B.5 by $\xi$. Since the Taylor expansion is performed at non-optimum points $(2n-1)\frac{\pi}{4}$, one iteration of the algorithm is not sufficient for adequate accuracy. In the second iteration, the function is expanded in the vicinity of the approximate solutions $y$ obtained in the first iteration. Table B.1 summarizes the results, achieved after first and second iterations for different values of $\xi$. As can be seen, after the second iteration a very high accuracy below $0.01\%$ with an exceptional case of 4.9% is achieved for the considered values of $\xi$. The calculated values are plotted in Fig. B.1 and Fig. B.2, indicating very good accuracy of the solutions for all considered values of $\xi$.

Table B.1: Approximate solutions of equation $\cot(x) = \xi x$

| $\xi = 0.1$ | | | | | | | | |
|---|---|---|---|---|---|---|---|---|
| n | $x_1$ | $0.1x_1$ | $\cot(x_1)$ | $\delta_1, \%$ | $x_2$ | $0.1x_2$ | $\cot(x_2)$ | $\delta_2, \%$ |
| 1 | 1.4285 | 0.14285 | 0.14326 | 0.29 | 1.42887 | 0.14289 | 0.14289 | $<10^{-6}$ |
| 2 | 4.29711 | 0.42971 | 0.44093 | 2.6 | 4.3058 | 0.43058 | 0.43058 | $<10^{-5}$ |
| 3 | 7.19652 | 0.71965 | 0.77205 | 7.28 | 7.22811 | 0.72281 | 0.72282 | $<10^{-4}$ |
| 4 | 10.13668 | 1.01367 | 1.15896 | 14.33 | 10.20025 | 1.02003 | 1.02005 | $<10^{-2}$ |

| $\xi = 1$ | | | | | | | | |
|---|---|---|---|---|---|---|---|---|
| n | $x_1$ | $x_1$ | $\cot(x_1)$ | $\delta_1, \%$ | $x_2$ | $x_2$ | $\cot(x_2)$ | $\delta_2, \%$ |
| 1 | 0.85931 | 0.85931 | 0.86212 | 0.33 | 0.86033 | 0.86033 | 0.86033 | $<10^{-4}$ |
| 2 | 3.38581 | 3.38581 | 4.01294 | 18.52 | 3.42559 | 3.42559 | 3.42602 | 0.01 |
| 3 | 6.38297 | 6.38297 | 9.98804 | 56.48 | 6.43729 | 6.43729 | 6.43774 | $<10^{-2}$ |
| 4 | 9.47726 | 9.47726 | 19.03667 | 100.87 | 9.52933 | 9.52933 | 9.52962 | $<10^{-2}$ |

| $\xi = 3$ | | | | | | | | |
|---|---|---|---|---|---|---|---|---|
| n | $x_1$ | $3x_1$ | $\cot(x_1)$ | $\delta_1, \%$ | $x_2$ | $3x_2$ | $\cot(x_2)$ | $\delta_2, \%$ |
| 1 | 0.41303 | 1.2391 | 2.28185 | 84.15 | 0.53486 | 1.60459 | 1.68786 | 4.93 |
| 2 | 3.24099 | 9.72298 | 10.02705 | 3.13 | 3.24399 | 9.73196 | 9.73197 | $<10^{-4}$ |
| 3 | 6.3357 | 19.00709 | 19.02649 | 0.1 | 6.33575 | 19.00725 | 19.00725 | $<10^{-8}$ |
| 4 | 9.45996 | 28.37989 | 28.40969 | 0.11 | 9.46 | 28.38 | 28.38 | $<10^{-9}$ |

# Bibliography

[1] S. Barker and G. M. Rebeiz, "Distributed MEMS true-time delay phase shifters and wide-band switches," *IEEE Transactions on Microwave Theory and Techniques*, vol. 46, no. 11, pp. 1881–1890, 1998.

[2] J. S. Hayden and G. M. Rebeiz, "Very low-loss distributed X-band and Ka-band MEMS phase shifters using metal-air-metal capacitors," *IEEE Transactions on Microwave Theory and Techniques*, vol. 51, no. 1, pp. 309–314, Jan. 2003.

[3] T. Vähä-Heikkilä and G. M. Rebeiz, "A 20-50 GHz reconfigurable matching network for power amplifier applications," in *Proc. IEEE MTT-S International Microwave Symposium Digest*, vol. 2, Jun. 6–11, 2004, pp. 717–720.

[4] J. Perruisseau-Carrier, *Microwave Periodic Structures Based on Microelectromechanical systems (MEMS) and Micromachining Techniques*. Ecole Polytechnique Federale de Lausanne: PhD Thesis, 2007.

[5] A. S. Nagra and R. A. York, "Distributed analog phase shifters with low insertion loss," *IEEE Transactions on Microwave Theory and Techniques*, vol. 47, no. 9, pp. 1705–1711, 1999.

[6] G. M. Rebeiz, *RF MEMS: theory, design, and technology*. John Wiley and Sons, 2003.

[7] L. Brillouin, *Wave Propagation in Periodic Structures*. McGraw-Hill Book Company, 1946.

[8] S. Lim, C. Caloz, and T. Itoh, "Metamaterial-based electronically controlled transmission-line structure as a novel leaky-wave antenna with tunable radiation angle and

beamwidth," *IEEE Transactions on Microwave Theory and Techniques*, vol. 53, no. 1, pp. 161–173, Jan. 2005.

[9] H. Kim, S.-J. Ho, M.-K. Choi, A. B. Kozyrev, and D. W. van der Weide, "Combined Left- and Right-Handed Tunable Transmission Lines With Tunable Passband and 0° Phase Shift," *IEEE Transactions on Microwave Theory and Techniques*, vol. 54, no. 12, pp. 4178–4184, Dec. 2006.

[10] C. Damm, M. Schussler, M. Oertel, and R. Jakoby, "Compact tunable periodically LC loaded microstrip line for phase shifting applications," in *Proc. IEEE MTT-S International Microwave Symposium Digest*, Long Beach, USA, Jun. 12–17, 2005.

[11] C. Damm, M. Schubler, J. Freese, and R. Jakoby, "Artificial line phase shifter with separately tunable phase and line impedance," in *Proc. 36th European Microwave Conference*, Manchester, UK, Sep. 10–15, 2006, pp. 423–426.

[12] S. Sheng, P. Wang, and C. K. Ong, "Compact Tunable Periodically LC Loaded Phase Shifter Using Left-Handed Transmission Line," *Microwave and Optical Technology Letters*, vol. 51, no. 9, pp. 2127–2129, Sept. 2009.

[13] C. Damm, J. Freese, M. Schussler, and R. Jakoby, "Electrically Controllable Artificial Transmission Line Transformer for Matching Purposes," *IEEE Transactions on Microwave Theory and Techniques*, vol. 55, no. 6, pp. 1348–1354, Jun. 2007.

[14] C.-Y. Kim, J. Yang, D.-W. Kim, and S. Hong, "A K-Band CMOS Voltage Controlled Delay Line Based on an Artificial Left-Handed Transmission Line," *IEEE Microwave and Wireless Components Letters*, vol. 18, no. 11, pp. 731–733, Nov. 2008.

[15] W. Tang and H. Kim, "Compact, Tunable Large Group Delay Line," *Microwave and Optical Technology Letters*, vol. 51, no. 12, pp. 2893–2895, Dec. 2009.

[16] Y. Wang, Y. Zhang, L. He, F. Liu, H. Li, and H. Chen, "Tunable Asymmetric Composite Right-/Left -Handed Transmission Line Directional Coupler Controlled by Applied Voltage," in *Asia-Pacific Microwave Conference*, Suzhou, China, 4-7 Dec. 2005.

[17] D. Kuylenstierna, A. Vorobiev, P. Linner, and S. Gevorgian, "Composite right/left handed transmission line phase shifter using ferroelectric varactors," *IEEE Microwave and Wireless Components Letters*, vol. 16, no. 4, pp. 167–169, 2006.

[18] X.-Y. Zhang, P. Wang, S. Sheng, Y. Ma, F. Xu, and C. K. Ong, "A novel structure for dc bias on varactors in composite right/left-handed transmission lines phase shifter using $Ba_{0.25}Sr_{0.75}TiO_3$ thin film," *IOP Journal of Physics D: Applied Physics*, vol. 42, p. 175103 (4pp), 2009.

[19] Y. Wang, M. J. Lancaster, F. Huang, P. M. Suherman, D. M. Holdom, and T. J. Jackson, "Superconducting Tunable Composite Right/Left-Handed Transmission Lines Using Ferroelectric Thin Films with a Resistive Bias Network," in *Proc. IEEE/MTT-S International Microwave Symposium*, Honolulu, USA, 3-8 June 2007, pp. 1415–1418.

[20] B. Lakshminarayanan and T. Weller, "MEMS phase shifters using cascaded slow-wave structures for improved impedance matching and/or phase shift," in *Proc. IEEE MTT-S International Microwave Symposium Digest*, vol. 2, Fort Worth, USA, Jun. 6–11, 2004, pp. 725–728.

[21] J. Saijets, P. Rantakari, M. Tuohiniemi, and T. Vähä-Heikkilä, "Comparison of dmtl phase shifter designs," in *10th International Symposium on RF MEMS and Microsystems MEMSWAVE*, Trento, Italy, June 2009, pp. 37–40.

[22] S. Balachandran, B. Lakshminarayanan, T. Weller, and M. Smith, "MEMS Tunable Planar Inductors using DC-Contact Switches," in *Proc. 34th European Microwave Conference*, vol. 2, Amsterdam, The Netherlands, Oct. 13, 2004, pp. 713–716.

[23] J.-W. Lin, C. C. Chen, and Y.-T. Cheng, "A robust high-Q micromachined RF inductor for RFIC applications," *IEEE Transactions on Electron Devices*, vol. 52, no. 7, pp. 1489–1496, 2005.

[24] N. A. Talwalkar, C. P. Yue, H. Gan, and S. S. Wong, "Integrated CMOS transmit-receive switch using LC-tuned substrate bias for 2.4-GHz and 5.2-GHz applications," *IEEE Journal of Solid-State Circuits*, vol. 39, no. 6, pp. 863–870, 2004.

[25] K.-H. Pao, C.-Y. Hsu, H.-R. Chuang, C.-L. Lu, and C.-Y. Chen, "A 3-10GHz Broadband CMOS T/R Switch for UWB Applications," in *Proc. 1st European Microwave Integrated Circuits Conf.*, Manchester, UK, Sept. 2006, pp. 452–455.

[26] Y. Jin and C. Nguyen, "Ultra-Compact High-Linearity High-Power Fully Integrated DC-20-GHz 0.18- μm CMOS T/R Switch," *IEEE Transactions on Microwave Theory and Techniques*, vol. 55, no. 1, pp. 30–36, 2007.

[27] S. Majumder, J. Lampen, R. Morrison, and J. Maciel, "A packaged, high-lifetime ohmic MEMS RF switch," in *Proc. IEEE MTT-S Int. Microwave Symp. Digest*, vol. 3, Philadelphia, Pennsylvania, June 8-13 2003, pp. 1935–1938.

[28] Y. Uno, K. Narise, T. Masuda, K. Inoue, Y. Adachi, K. Hosoya, T. Seki, and F. Sato, "Development of SPDT-structured RF MEMS switch," in *Proc. Int. Solid-State Sensors, Actuators and Microsystems Conf. TRANSDUCERS*, Denver, Colorado, USA, June 21 - 25 2009, pp. 541–544.

[29] C. Siegel, V. Ziegler, U. Prechtel, B. Schonlinner, and H. Schumacher, "Very low complexity RF-MEMS technology for wide range tunable microwave filters," in *Proc. European Microwave Conference*, vol. 1, Paris, France, Oct. 4–6, 2005, p. 4pp.

[30] C. Siegel, V. Ziegler, B. Schonlinner, U. Prechtel, and H. Schumacher, "Simplified RF-MEMS Switches Using Implanted Conductors and Thermal Oxide," in *Proc. 36th European Microwave Conference*, Manchester, UK, Sep. 10–15, 2006, pp. 1735–1738.

[31] C. Siegel, V. Ziegler, B. Schönlinner, U. Prechtel, and H. Schumacher, "RF-MEMS based 2-bit reflective phase shifter at X-Band for reconfigurable reflect-array antennas," in *8th International Symposium on RF MEMS and Microsystems MEMSWAVE*, Barcelona, Spain, June 2007.

[32] J. A. Pelesko and D. H. Bernstein, *Modeling MEMS and NEMS*. Chapman&Hall/CRC, 2003.

[33] D. Neculoiu, R. Plana, P. Pons, A. Muller, D. Vasilache, I. Petrini, C. Buiculescu, and P. Blondy, "Microwave characterization of membrane supported coplanar waveguide transmission lines - electromagnetic simulation and experimental results," in *Proc. Int. Semiconductor Conf. CAS 2001*, vol. 1, Sinaia, Romania, 09-13 Oct. 2001, pp. 151–154.

[34] D. Neculoiu, P. Pons, M. Saadaoui, L. Bary, D. Vasilache, K. Grenier, D. Dubuc, A. Muller, and R. Plana, "Membrane supported Yagi-Uda antennae for millimetre-wave applications," *IEE Proceedings -Microwaves, Antennas and Propagation*, vol. 151, no. 4, pp. 311–314, 2004.

[35] J. Kusterer, F. Hernandez, S. Haroon, P. Schmid, A. Munding, R. Müller, and E. Kohn, "Bi-stable micro actuator based on stress engineered nano-diamond," *Diamond & Related Materials, Elsevier*, vol. 15, pp. 773–6, 2006.

[36] H. C. Nathanson, W. E. Newell, R. A. Wickstrom, and J. Davis, J. R., "The resonant gate transistor," *IEEE Transactions on Electron Devices*, vol. 14, no. 3, pp. 117–133, 1967.

[37] D. Elata and H. Bamberger, "A lower bound for the dynamic pull-in of electrostatic actuators," in *Proc. Eur. Micro Nano Syst.*, Paris, France, 2004, p. pp. 8385.

[38] D. Elata and D. Bamberger, "On the dynamic pull-in of electrostatic actuators with multiple degrees of freedom and multiple voltage sources," *Journal of Microelectromechanical Systems*, vol. 15, no. 1, pp. 131–140, 2006.

[39] V. Leus and D. Elata, "On the Dynamic Response of Electrostatic MEMS Switches," *Journal of Microelectromechanical Systems*, vol. 17, no. 1, pp. 236–243, 2008.

[40] R. K. Gupta and S. D. Senturia, "Pull-in time dynamics as a measure of absolute pressure," in *Proc. , IEEE. Workshop Tenth Annual Int Micro Electro Mechanical Systems MEMS '97*, Nagoya, Japan, 26-30 Jan. 1997, pp. 290–294.

[41] IHP Microelectronics. [Online]. Available: http://www.ihp-microelectronics.com/

[42] R. E. Collin, *Foundations for Microwave Engineering*, 2nd ed.   IEEE Press, 2001.

[43] T. Lisec, C. Huth, M. Shakhray, and B. Wagner, "Surface-micromachined capacitive RF switches with high thermal stahility and low drift using Ni as structural material," in *Proc. MEMSWAVE 2004*, Uppsala, June 2004, pp. C33–C36.

[44] J. Ruan, N. Nolhier, M. Barfleur, L. Bary, N. Mauran, F. Coccetti, T. Lisec, and R. Plana, "Dielectric Material Charging and ESD Stress of AlN-based Capacitive RF MEMS," in *8th*

*International Symposium on RFMEMS and RF Microsystems MEMSWAVE*, Barcelona, Spain, 2007, pp. 179–182.

[45] T. Lisec, C. Huth, and B. Wagner, "Dielectric material impact on capacitive RF MEMS reliability," in *Proc. 34th European Microwave Conf*, vol. 1, Amsterdam, the Netherlands, 11-15 Oct. 2004, pp. 73–76.

[46] M. Fernandez-Bolanos, T. Lisec, P. Dainesi, and A. M. Ionescu, "Thermally Stable Distributed MEMS Phase Shifter for Airborne and Space Applications," in *Proc. 38th European Microwave Conf. EuMC 2008*, Amsterdam, the Netherlands, 27-31 Oct. 2008, pp. 100–103.

[47] H. Gamble, B. Armstrong, S. Mitchell, Y. Wu, V. Fusco, and J. Stewart, "Low-loss CPW lines on surface stabilized high-resistivity silicon," *Microwave and Guided Wave Letters, IEEE*, vol. 9, no. 10, pp. 395 – 397, Oct 1999.

[48] E. Valletta, J. V. Beek, A. D. Dekker, N. Pulsford, H. F. F. Jos, L. de Vreede, L. K. Nanver, and J. N. Burghartz, "Design and Characterization of Integrated Passive Elements on High Ohmic Silicon," in *Microwave Symposium Digest, 2003 IEEE MTT-S International*, vol. 2, Philadelphia, Pennsylvania, USA, June 8-13 2003, pp. 1235 – 1238.

[49] T. Feger, T. Purtova, T. Lisec, N. Pour Aryan, C. Chu, and H. Schumacher, "Influence of Oxidized High-Resistivity Silicon on the Loss and Phase Velocity of CPW RFMEMS," in *11th International Symposium on RF MEMS and RF Microsystems MEMSWAVE*, Otranto, June 2010.

[50] D. Peroulis, S. Pacheco, K. Sarabandi, and P. B. Katehi, "Mems devices for high isolation switching and tunable filtering," in *Proc. IEEE MTT-S Int. Microwave Symp. Digest*, vol. 2, Boston, Massachusetts, USA, 11.16 June 2000, pp. 1217–1220.

[51] F. Giacomozzi, *FBK RF-MEMS process description*.

[52] P. Farinelli, A. Ocera, B. Margesin, F. Giacomozzi, and R. Sorrentino, "High Performance RF-MEMS cantilever switch," in *7th International Symposium on RF MEMS and RF Microsystems MEMSWAVE 2006*, Orvieto, Italy, 27-30 June 2006.

[53] E. K. I. Hamad, A. M. E. Safwat, and A. S. Omar, "Controlled capacitance and inductance behaviour of l-shaped defected ground structure for coplanar waveguide," *IEE Proceedings -Microwaves, Antennas and Propagation*, vol. 152, no. 5, pp. 299–304, 2005.

[54] A. Gorur, C. Karpuz, and M. Alkan, "Characteristics of periodically loaded cpw structures," *IEEE Microwave and Guided Wave Letters*, vol. 8, no. 8, pp. 278–280, 1998.

[55] M. J. W. Rodwell, S. T. Allen, R. Y. Yu, M. G. Case, U. Bhattacharya, M. Reddy, E. Carman, M. Kamegawa, Y. Konishi, J. Pusl, and R. Pullela, "Active and nonlinear wave propagation devices in ultrafast electronics and optoelectronics," *Proceedings of the IEEE*, vol. 82, no. 7, pp. 1037–1059, 1994.

[56] C. Elachi, "Waves in active and passive periodic structures: A review," *Proceedings of the IEEE*, vol. 64, no. 12, pp. 1666–1698, Dec. 1976.

[57] T. Tamir, H. C. Wang, and A. A. Oliner, "Wave Propagation in Sinusoidally Stratified Dielectric Media," *IEEE Transactions on Microwave Theory and Techniques*, vol. 12, no. 3, pp. 323–335, 1964.

[58] N. W. McLachlan, *Theory and Application of Mathieu Functions*. New York : Dover Publ., 1964.

[59] F. E. Gardiol, *Lossy Transmission Lines*. Artech House, 1987.

[60] G. Strang, *Linear Algebra and Its Applications*. Harcourth Brace Jovanovich, 1988.

[61] A. S. Nagra, J. Xu, E. Erker, and R. A. York, "Monolithic gaas phase shifter circuit with low insertion loss and continuous 0-360&deg; phase shift at 20 ghz," *IEEE Microwave and Guided Wave Letters*, vol. 9, no. 1, pp. 31–33, 1999.

[62] L. Zhu, "Guided-wave characteristics of periodic coplanar waveguides with inductive loading - unit-length transmission parameters," *IEEE Transactions on Microwave Theory and Techniques*, vol. 51, no. 10, pp. 2133–2138, 2003.

[63] J. H. Schaffner, D. F. Sievenpiper, R. Y. Loo, J. J. Lee, and S. W. Livingston, "A wideband beam switching antenna using rf mems switches," in *Proc. IEEE Antennas and Propagation Society Int. Symp*, vol. 3, Boston, MA, USA, 08-13 Jul. 2001, pp. 658–661.

[64] H. Legay, B. Pinte, M. Charrier, A. Ziaei, E. Girard, and R. Gillard, "A steerable reflectarray antenna with mems controls," in *Proc. IEEE Int Phased Array Systems and Technology Symp*, Boston, Massachusetts, USA, 14-17 Oct. 2003, pp. 494–499.

[65] J. Huang and R. J. Pogorzelski, "A ka-band microstrip reflectarray with elements having variable rotation angles," *IEEE Transactions on Antennas and Propagation*, vol. 46, no. 5, pp. 650–656, 1998.

[66] B. Mencagli, R. V. Gatti, L. Marcaccioli, and R. Sorrentino, "Design of large mm-wave beam-scanning reflectarrays," in *Proc. European Microwave Conf*, vol. 3, Paris, France, 4-6 Oct. 2005.

[67] L. Marcaccioli, B. Mencagli, R. V. Gatti, T. Feger, T. Purtova, H. Schumacher, and R. Sorrentino, "Beam Steering MEMS mm-Wave Reflectarrays," in *7th International Symposium on RF MEMS and Microsystems MEMSWAVE*, Orvieto, Italy, Jun. 27–30, 2006.

[68] T. Vähä-Heikkilä and M. Ylonen, "G-Band Distributed Microelectromechanical Components Based on CMOS Compatible Fabrication," *IEEE Transactions on Microwave Theory and Techniques*, vol. 56, no. 3, pp. 720–728, Mar. 2008.

[69] T. Vähä-Heikkilä, *Design and Modeling Issues for High Power Handling RF MEMS*, ser. RF MEMS Industrial Workshop.   Fodele, Greece: Workshop, 1 July 2008.

[70] R. Malmqvist, C. Samuelsson, P. Rantakari, P. Frijlink, D. Smith, W. Simon, J. Saijets, T. Vaha-Heikkila, and R. Baggen, "RF MEMS-MMIC building blocks for emerging wireless systems and RF-sensing applications," in *Proc. European Microwave Integrated Circuits Conf. EuMIC 2009*, Rome, Italy, 28-29 Sept. 2009, pp. 363–366.

[71] W. Simon, L. Baggen, and D. Smith, "Innovative RF MEMS switches on GaAs," in *Proc. 14th Int Antenna Technology and Applied Electromagnetics & the American Electromagnetics Conf. (ANTEM-AMEREM) Symp*, Ottawa, ON, 5-8 July 2010, pp. 1–4.

[72] P. Rantakari, R. Malmqvist, C. Samuelsson, R. Leblanc, D. Smith, R. Jonsson, W. Simon, J. Saijets, R. Baggen, and T. Vähä-Heikkilä, "Wide-band radio frequency micro electro-mechanical systems switches and switching networks using a gallium arsenide

monolithic microwave-integrated circuits foundry process technology," *IET Microwaves, Antennas & Propagation*, vol. 5, no. 8, pp. 948–955, 2011.

[73] F. Crispoldi, A. Pantellini, S. Lavanga, A. Nanni, P. Romanini, L. Rizzi, P. Farinelli, and C. Lanzieri, "New fabrication process to manufacture RF-MEMS and HEMT on GaN/Si substrate," in *Proc. European Microwave Integrated Circuits Conf. EuMIC 2009*, Rome, Italy, 28-29 Sept. 2009, pp. 387–390.

[74] M. Kaynak, K. E. Ehwald, J. Drews, R. Scholz, F. Korndorfer, D. Knoll, B. Tillack, R. Barth, M. Birkholz, K. Schulz, Y. M. Sun, D. Wolansky, S. Leidich, S. Kurth, and Y. Gurbuz, "BEOL embedded RF-MEMS switch for mm-wave applications," in *Proc. IEEE Int. Electron Devices Meeting (IEDM)*, Baltimore, MD, USA, 7-9 Dec. 2009, pp. 1–4.

[75] M. Kaynak, M. Wietstruck, R. Scholz, J. Drews, R. Barth, K. E. Ehwald, A. Fox, U. Haak, D. Knoll, F. Korndorfer, S. Marschmeyer, K. Schulz, C. Wipf, D. Wolansky, B. Tillack, K. Zoschke, T. Fischer, Y. S. Kim, J. S. Kim, W.-G. Lee, and J. W. Kim, "BiCMOS embedded RF-MEMS switch for above 90 GHz applications using backside integration technique," in *Proc. IEEE Int. Electron Devices Meeting (IEDM)*, San Francisco, CA, USA, 6-8 Dec. 2010.

[76] M. Kaynak, K. E. Ehwald, R. Scholz, F. Korndorfer, C. Wipf, Y. M. Sun, B. Tillack, S. Zihir, and Y. Gurbuz, "Characterization of an embedded RF-MEMS switch," in *Proc. Topical Meeting Silicon Monolithic Integrated Circuits in RF Systems (SiRF)*, New Orleans, LA, USA, 11-13 Jan. 2010, pp. 144–147.

[77] M. Kaynak, K. E. Ehwald, J. Drews, R. Scholz, F. Korndörfer, C. Wipf, D. Knoll, R. Barth, M. Birkholz, K. Schulz, D. Wolansky, and B. Tillack, "Embedded MEMS modules for BiCMOS process," in *Proc. German Microwave Conf*, Berlin, Germany, 15-17 March 2010, pp. 78–81.

[78] M. Kaynak, F. Korndörfer, M. Wietstruck, D. Knoll, R. Scholz, C. Wipf, C. Krause, and B. Tillack, "Robustness and reliability of BiCMOS embedded RF-MEMS switch," in *Proc. IEEE 11th Topical Meeting Silicon Monolithic Integrated Circuits in RF Systems (SiRF)*, Phoenix, AZ, USA, 17-19 Jan. 2011, pp. 177–180.

[79] V. Ziegler, C. Siegel, B. Schonlinner, U. Prechtel, and H. Schumacher, "RF-MEMS switches based on a low-complexity technology and related aspects of MMIC integration," in *Proc. European Gallium Arsenide and Other Semiconductor Application Symposium GAAS 2005*, Paris, France, 3-4 Oct. 2005, pp. 289–292.

[80] M. A. Qureshi, T. Purtova, I. Kallfass, A. Leuther, T. Feger, and H. Schumacher, "G and V-band RF MEMS Switches and G-band SPDT Monolithically Integrated in m-HEMT Technology," in *Second IEEE Germany Student Conference*, Hamburg, Germany, May 2010.

[81] S. Frankel, "Complete Approximate Solutions of the Equation $x = \tan(x)$," *National Mathematics Magazine*, vol. 11, no. 4, pp. 177–182, 1937.

# Publications

[1] T. Purtova and H. Schumacher. *Handbook of MEMS for wireless and mobile applications, Ch. Overview of RF MEMS technology and applications*. Woodhead Publlishing, in preparation.

[2] A. C. Ulusoy, M. Kaynak, T. Purtova, B. Tillack, and H. Schumacher. A 60 to 77GHz Switchable LNA in an RF-MEMS Embedded BiCMOS Technology. *Microwave and Wireless Components Letters, IEEE*, 22:430 –432, Aug. 2012.

[3] M. Kaynak, M. Wietstruck, W. Zhang, J. Drews, R. Barth, D. Knoll, F. Korndoerfer, R. Scholz, K. Schulz, C. Wipf, B. Tillack, K. Kaletta, K. Zoschke, M. Wilke, O. Ehrmann, T. Purtova, A. C. Ulusoy, G. Liu, H. Schumacher, and M. Suchodoletz. Packaged BiCMOS Embedded RF-MEMS Switches with Integrated Inductive Loads. In *accepted to the IEEE MTT-S International Microwave Symposium*, 2012.

[4] M. Kaynak, M. Wietstruck, W. Zhang, J. Drews, R. Scholz, D. Knoll, F. Korndörfer, C. Wipf, K. Schulz, M. Elkhouly, K. Kaletta, M. v. Suchodoletz, K. Zoschke, M. Wilke, O. Ehrmann, V. Mühlhaus, G. Liu, T. Purtova, A. C. Ulusoy, H. Schumacher, and B. Tillack. RF-MEMS Switch Module in a 0.25 µm BiCMOS Technology. In *12th Topical Meeting on Silicon Monolithic Integrated Circuits in RF Systems (SiRF 2012)*, Santa Clara, CA, USA, 16-18 January 2012.

[5] Gang Liu, Mehmet Kaynak, Tatyana Purtova, A. Cagri Ulusoy, Bernd Tillack, and Hermann Schumacher. Dual-Band Millimeter-Wave VCO with Embedded RF-MEMS Switch Module in BiCMOS Technology. In *12th Topical Meeting on Silicon Monolithic Integrated Circuits in RF Systems (SiRF 2012)*, Santa Clara, CA, USA, 16-18 January 2012.

[6] T. Bartnitzek, T. Purtova, C. Rusch, S. Kaminski, and T. Feger. An Investigation of the Process Stability of RF SiP Made of DuPont 943 and 9K7. In *IMAPS/ACerS 7$^{th}$ International Conference and Exhibition*, San Diego, USA, Apr. 2011.

[7] T. Bartnitzek, T. Purtova, C. Rusch, S. Kaminski, and T. Feger. An Investigation of the Process Stability of RF SiP Made of DuPont 943 and 9K7. *Journal of Microelectronics and Electronic Packaging*, 8(1):34–41, 2011.

[8] T. Feger, T. Purtova, Th. Lisec, N. Pour Aryan, C. Chu, and H. Schumacher. Influence of Oxidized High-Resistivity Silicon on the Loss and Phase Velocity of CPW RFMEMS. In *11th International Symposium on RF MEMS and RF Microsystems MEMSWAVE 2010*, Otranto, Italy, June 2010.

[9] M. Z. Alam, T. Purtova, G. Liu, M. Kaynak, and H. Schumacher. V/W-band Wide Tuning Range VCO with BEOL Embedded RF-MEMS in BiCMOS Technology. In *Proc. Second IEEE Germany Student Conference*, Hamburg, Germany, May 2010.

[10] M. A. Qureshi, T. Purtova, I. Kallfass, A. Leuther, T. Feger, and H. Schumacher. G and V-band RF MEMS Switches and G-band SPDT Monolithically Integrated in m-HEMT Technology. In *Proc. Second IEEE Germany Student Conference*, Hamburg, Germany, May 2010.

[11] G. Liu, B. Schleicher, T. Purtova, and H. Schumacher. Fully Integrated Millimeter-Wave VCO Using Slow-Wave Thin-Film Microstrip Lines for Chip Size Reduction. In *Proc. 4th European Microwave Integrated Circuits Conference*, pages 116–119, Sept. 28–29, 2009.

[12] T. Feger, T. Purtova, T. Lisec, and H. Schumacher. A 35 GHz Single-Pole-Double-Throw Using Optimised Shunt Capacitive RF-MEMS Switches. In *Proc. 10th International Symposium on RF MEMS and Microsystems MEMSWAVE*, Trento, Italy, July 7–8, 2009.

[13] T. Bartnitzek, B. Schönlinner, W. Gautier, S. Cheng, A. Rydberg, T. Purtova, G. Qu, T. Feger, J. Janes, T. Vähä-Heikkilä, and A. Ziaei. Ceramic systems in package for RF and microwave. In *IMAPS Advanced Technology Workshop on RF and Microwave Packaging*, San Diego, USA, Oct. 2008.

[14] T. Purtova, T. Feger, T. Lisec, B. Wagner, and H. Schumacher. Compact Reflection-type RF-MEMS Phase Shifter for Beam-Steerable Reflectarrays. In *Proc. 9th International Symposium on RF MEMS and Microsystems MEMSWAVE*, Fodele, Greece, July 2–3, 2008.

[15] T. Purtova, T. Vähä-Heikkilä, T. Feger, T. Lisec, J. Janes, B. Wagner, and H. Schumacher. One Single DMTL-based Circuit for RF-MEMS Impedance Tuners and Phase Shifters. In *Proc. 9th International Symposium on RF MEMS and Microsystems MEMSWAVE*, Fodele, Greece, July 2–3, 2008.

[16] S. Chartier, B. Schleicher, T. Feger, T Purtova, G. Fischer, and H. Schumacher. 79 GHz fully integrated fully differential Si/SiGe HBT amplifier for automitive radar applicatioins. *Analog Integrated Circuits and Signal Processing, Springer*, 55(1):77–83, 2008.

[17] S. Chartier, B. Schleicher, T. Feger, T. Purtova, and H. Schumacher. 79 GHz Fully Integrated Fully Differential Si/SiGe HBT Amplifier for Automotive Radar Applications. In *Proc. 13th IEEE International Conference on Electronics, Circuits and Systems ICECS '06*, pages 1011–1014, December 10–13, 2006.

[18] L. Marcaccioli, B. Mencagli, R. Vincenti Gatti, T. Feger, T. Purtova, H. Schumacher, and R. Sorrentino. Beam Steering MEMS mm-Wave Reflectarrays. In *Proc. 7th International Symposium on RF MEMS and Microsystems MEMSWAVE*, Orvieto, Italy, June 27–30, 2006.

[19] I. Kallfass, T. Purtova, H. Schumacher, A. Brokmeier, and W. Ludwig. One single traveling-wave MMIC for highly linear broadband mixers and variable gain amplifiers. In *Proc. IEEE MTT-S International Microwave Symposium Digest*, June 12–17, 2005.

## i want morebooks!

Buy your books fast and straightforward online - at one of world's fastest growing online book stores! Environmentally sound due to Print-on-Demand technologies.

## Buy your books online at
## www.get-morebooks.com

Kaufen Sie Ihre Bücher schnell und unkompliziert online – auf einer der am schnellsten wachsenden Buchhandelsplattformen weltweit! Dank Print-On-Demand umwelt- und ressourcenschonend produziert.

## Bücher schneller online kaufen
## www.morebooks.de

VDM Verlagsservicegesellschaft mbH
Heinrich-Böcking-Str. 6-8
D - 66121 Saarbrücken

Telefon: +49 681 3720 174
Telefax: +49 681 3720 1749

info@vdm-vsg.de
www.vdm-vsg.de

Printed by Books on Demand GmbH, Norderstedt / Germany